THE REAL JAMES BOND

―― *Jim Wright* ――

A True Story of Identity Theft, Avian Intrigue & Ian Fleming

SCHIFFER PUBLISHING

4880 Lower Valley Road · Atglen, PA 19310

Other Schiffer Books by Jim Wright:
The Nature of the Meadowlands, ISBN 978-0-7643-4186-1

Copyright © 2020 by Jim Wright
Library of Congress Control Number: 2019946905

All rights reserved. No part of this work may be reproduced or used in any form or by any means—graphic, electronic, or mechanical, including photocopying or information storage and retrieval systems—without written permission from the publisher.

The scanning, uploading, and distribution of this book or any part thereof via the Internet or any other means without the permission of the publisher is illegal and punishable by law. Please purchase only authorized editions and do not participate in or encourage the electronic piracy of copyrighted materials.

"Schiffer," "Schiffer Publishing, Ltd.," and the pen and inkwell logo are registered trademarks of Schiffer Publishing, Ltd.

Designed by Jack Chappell
Cover design by Molly Shields
Type set in Caviar Dreams/Forque/Minion Pro
Author photo by Kevin Watson

ISBN: 978-0-7643-5902-6
Printed in China

Published by Schiffer Publishing, Ltd.
4880 Lower Valley Road
Atglen, PA 19310
Phone: (610) 593-1777; Fax: (610) 593-2002
E-mail: Info@schifferbooks.com
Web: www.schifferbooks.com

For our complete selection of fine books on this and related subjects, please visit our website at www.schifferbooks.com. You may also write for a free catalog.

Schiffer Publishing's titles are available at special discounts for bulk purchases for sales promotions or premiums. Special editions, including personalized covers, corporate imprints, and excerpts, can be created in large quantities for special needs. For more information, contact the publisher.

We are always looking for people to write books on new and related subjects. If you have an idea for a book, please contact us at proposals@schifferbooks.com.

For Elby, Corie, Patty, Lily, and Lucille

CONTENTS

Acknowledgments	005
Introduction: A Spy Is Born	007
Chapter 001. Encounter at Goldeneye	013
Chapter 002. The Bonds of Philadelphia	021
Chapter 003. Have Gun, Will Travel	033
Chapter 004. Shotguns & Arsenic	045
Chapter 005. *Birds of the West Indies*	057
Chapter 006. The Bond-Fleming Bond	065
Chapter 007. *You Only Live Twice*	081
Chapter 008. Twitchers & Spooks	087
Chapter 009. Was Jim Bond a Spy?	101
Chapter 010. Bond's Legacy	107
Appendix 001. In Bond's Footsteps	112
Appendix 002. Goldfincher	127
Bibliography	133
Index	140

Acknowledgments

First, a huge thank-you to my wife, Patty Finn, who offered much-appreciated advice and a lot of patience for the two and a half years I worked on this book. LYS.

I also want to thank the following for their help, beginning with David Contosta, the author and Chestnut Hill College professor. A friend of Jim and Mary Bond, David wrote *The Private Life of James Bond* in 1993 and graciously shared his knowledge and archives with me.

In Philadelphia and environs, I wish to thank Nate Rice, Robert McCracken Peck, Jennifer Vess, Jason D. Weckstein, Dan Thomas, and Carolyn Belardo of the Academy of Natural Sciences of Drexel University; Caitlin Goodman and Joseph Shemtov of the Free Library of Philadelphia; Deanne D'Emilio, Jerry McLaughlin, Sister Kati McMahon, and Dawn Caruano of Gwynedd Mercy University; George Armistead; Harry Armistead; Frank Gill; Bert Filmyer of the Delaware Valley Ornithological Club; Scott McConnell; Keith S. Thomson; and Sue Phillips of the Church of the Messiah.

In Cuba: Orlando Garrido, Gary Markowski of the Caribbean Conservation Trust, and Frank Medina and Osmani Borrego Fernandez. In Jamaica: Ricardo Miller of Arrowhead Birding Tours, Catherine Levy of the Windsor Research Centre, Anthony Gambrill, and Ramsey Acosta, Lauren Wintemberg, and the staff at GoldenEye.

In Maine: Michael J. Good of Downeast Nature Tours, Kate Chaplin, Becky Marvil of the Acadia Bird Festival, Sean Charette of the Wendell Gilley Museum, Rich MacDonald of the Natural History Center, Elly Andrews and Kate Young of the Northeast Library, and Karen Craig of the Southwest Harbor Historical Society.

Elsewhere: David Levesque of St. Paul's School, Concord, New Hampshire; Mike Berry of the University of South Carolina; Mark Ridgway and Tace Fox of Harrow School, Harrow, London; Jonathan Smith of Trinity College, Cambridge, England; Anna Raff; Samuel Turvey; Storrs Olson; Herbert Raffaele; Eamonn Cummings; the staff of Lee Memorial Library in Allendale, New Jersey; Hiroki Fukuda; Charles Tyson Jr.; and Mimi Sabatino (for her help and expertise on the images in this book).

Jim Bond in the early 1950s. *Free Library of Philadelphia, Rare Book Department*

INTRODUCTION

A Spy Is Born

The name of a man is a numbing blow from which he never recovers.

—Marshall McLuhan, *Understanding Media*

This is the story of the most notorious case of identity theft in history.

Long before Ian Lancaster Fleming became a bestselling author, a single-minded Philadelphia ornithologist named James Bond wrote *Birds of the West Indies*, based on repeated expeditions to the Bahamas and the Caribbean from 1927 to 1935. The 480-page book, published in 1936, featured black-and-white line drawings of birds by noted artist Earl Poole. It quickly became the must-have book on the subject.

Fast forward to early 1952. Fleming was starting to flesh out his maiden 007 novel, *Casino Royale*, and needed a name for his imaginary secret agent. As he would often do when inventing characters for his books, Fleming simply lifted a name from real life. He looked at the distinctive white dust jacket of *Birds of the West Indies* on his bookshelf, noted the author's name in capital letters, and took it: James Bond. Period.

No middle name. Not even a middle initial.

Just nine purloined letters.

A spy was born.

Along with Sherlock Holmes and Harry Potter, British secret agent James Bond is one of the most popular fictional characters in the history of both publishing and moviemaking. Joined by an array of femme fatales and such villains as Dr. No, Mr. Big, and Auric Goldfinger, Ian Fleming's legendary 007 continues to flourish nearly seventy years after the British thriller writer first put pen to paper at Goldeneye, his winter home in Jamaica.

A key part of Bond lore is how and why Ian Fleming stole the naturalist/author's name for his fictional spy. Since then, the tale of Ian Fleming and the real James Bond has developed a life of its own, gradually accumulating embellishments and inaccuracies—beginning with the oft-repeated assertion that Bond gave Fleming permission to take his name and that Bond enjoyed the fame.

In fact, Bond grew to hate the 007 connection, even as his wife, Mary, did her best to promote it. And although the appropriation of his identity caught Bond by surprise, there is a long tradition of ornithologist spies—including at least seven of his contemporaries. Some people have even speculated that the real James Bond was a spy as well.

The mythology of James Bond will no doubt continue to grow, but its origins should be rooted in fact. Today, the genuine Bond is known mostly as an asterisk to the 007 phenomenon. He has become fodder for trivia questions ("What was the first film to reference ornithologist James Bond, whom Ian Fleming named the character after?") and crosswords, such as this clue from the January 10, 2016, *New York Times* puzzle: Across, 42: "Ornithologist James of whom Ian Fleming was a fan."

Bond deserves better. From the 1920s through the 1960s, he took more than one hundred scientific expeditions to the West Indies. These were the days before jet airliners and online booking services, and he would travel to the Caribbean for up to nine months at a time, living on a shoestring. The seasick-prone Bond went from

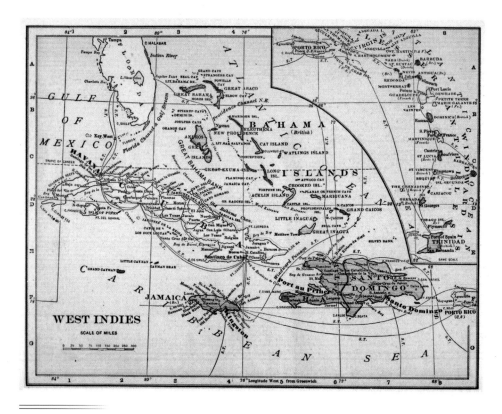

Map of the West Indies, ca. 1911. *Author's collection*

island to island on tramp steamers, canoes, and primitive sailboats in his efforts to collect and describe specimens of the birds on each island.

Bond's fieldwork resulted in his celebrated 1936 book *Birds of the West Indies*—one of the first of its kind. It helped introduce the world to such exotic creatures as Cuba's bee hummingbird (the world's smallest bird), the rare Cuban ivory-billed woodpecker, and the breathtaking red-billed streamertail (the national bird of Jamaica). The book remained in print in various editions for an incredible seven decades.

His research also resulted in his landmark zoogeographical theory in 1934 that the birds of the Caribbean were most closely related to North American birds, not South American birds, as had previously been thought. He supported his theory in dozens of scientific papers over the years.

This conclusion eventually led the noted evolutionary biologist David Lack to propose, in 1973, that the name "Bond's Line" be used to denote this boundary between the Lesser Antilles and Tobago. A quarter century earlier, Lack had written *Darwin's Finches*, forever linking Charles Darwin with the thirteen species of finches in the Galápagos.

Similarly, Bond discovered a geographical barrier that had once run across part of the island of Hispaniola, home to the Dominican Republic and Haiti. Bond found that

Front cover of *Birds of the West Indies*, 1936 edition. *Courtesy of Jack Holloway*

James Bond's *Birds of the West Indies* helped introduce the world to the bee hummingbird—the smallest bird on the planet. *Photo by author*

Introduction

The red-billed streamertail, one of Ian Fleming's favorite hummingbirds, is the national bird of Jamaica and part of the logo for ianfleming.com. *Photo by author*

Samuel Turvey and his team from the Zoological Society of London found a new species of hutia in Haiti and named it after Jim Bond—*Plagiodontia aedium bondi*. *Courtesy of J. Nunez-Mino, the Last Survivors Project*

the animals are noticeably different on either side of the invisible barrier. It turns out that a shallow sea channel prevented animals from moving freely across the island many centuries ago. That barrier is also known as Bond's Line.

Long before the 007 connection, Bond was a conservationist as well as an ornithologist, campaigning for increased protections for West Indies bird populations.

In the introduction to *Birds of the West Indies*, Bond wrote: "In no other part of the world . . . are so many birds in danger of extinction as in the West Indies, and at least twelve forms have become extinct in the last hundred years. It is to be hoped that the island authorities will show more concern for the welfare of their birds so there may yet be a possibility to save the rare species from being annihilated. Bird sanctuaries should be created where no hunting of any kind is permitted and there should be a complete ban on commercial collecting."

Although Bond appeared to be a genteel, pipe-smoking naturalist, he was an adventurer with a reputation for occasional ruthlessness. An expert marksman, he saw birds the same way John James Audubon did in days of yore—down the barrel of a gun.

Over a period of four decades, Bond collected more than 290 of the 300 bird species known to the West Indies. Birds, fish, and insects that he found on those islands are in the collections of such museums as the Academy of Natural Sciences of Philadelphia (ANSP), New York's American Museum of Natural History, the Smithsonian, Yale's Peabody Museum of Natural History, and Harvard's Museum of Comparative Zoology.

In recognition of Bond's achievements, researchers have named a new family of Caribbean plants, several fish, a grasshopper, a stink bug, and a subspecies of barn owl and seven other birds in his honor.

After British scientist Samuel Turvey's research team in Haiti discovered a previously unknown species of a mammal called a hutia in Haiti in 2015, they named it *Plagiodontia aedium bondi*. "James Bond was the first naturalist to notice that there was this discontinuity in distributions, which reflects an ancient marine barrier," Turvey said in an interview. "Since then, that boundary has been known as Bond's Line, and that marked one of the hutia distributions as well."

Naming the rodent after Bond served another purpose—publicity. "It allowed us to come up with various puns based on various James Bond film titles and quotes, which helps get the word out about a species which otherwise would probably not attract much attention or interest," Turvey said, adding that the media attention also underscores environmental issues that mammals face on Hispaniola.

The real James Bond appeared on two 2014 postage stamps in the West African nation of Mali. And at least one Bond movie—*Die Another Day*—paid homage to the real Bond when Pierce Brosnan's 007 posed as an ornithologist who carried a later edition of *Birds of the West Indies* and eyed a "bird" named Jinx (Halle Berry) in Cuba.

A world traveler for much of his life, Bond was forced to narrow his range in December 1974, when he had surgery for prostate cancer. He later developed leukemia and died on Valentine's Day 1989.

The Republic of Mali issued two postage stamps honoring Bond, one depicting Bond ca. 1965 and the other in his later years. *Author's collection*

An aerial view of the GoldenEye resort on the north shore of Jamaica. *Mark Collins photo, courtesy of Island Outpost*

CHAPTER 001

Encounter at Goldeneye

Goldeneye, Ian Fleming's legendary retreat in Jamaica, sits atop a bluff with commanding views of a cerulean-blue Caribbean. The former British naval intelligence officer first visited the island with his lifelong buddy, Ivar Bryce, when they participated in a U-boat conference during WWII, and Fleming immediately trained his sights on returning.

As Bryce recounted in his memoir, *You Only Live Once*, Fleming told him: "I have made a great decision. When we have won this blasted war, I am going to live in Jamaica. Just live in Jamaica and lap it up, and swim in the sea and write books. That is what I want to do, and I want your help, as you will probably get out of the war before I can. You must find the right bit of Jamaica for me to buy. Ten acres or so, away from the towns and on the coast. Find the perfect place; I want it signed and sealed. Please help."

After the war, Bryce got to work. The first time he set eyes on a secluded cliffside property on the north coast, he knew he'd found Fleming's perfect place. Bryce bought the 14-acre property for 2,000 pounds—about $8,300—on the spot.

In 1946, Fleming set out to build a modest one-story house on the site, a former donkey race course. As Matthew Parker recounted in his 2015 biography *Goldeneye: Where Bond Was Born*, the area was also known as Rock Edge and Rotten Egg Bay. Fleming specified that the three-bedroom bungalow be simple, and that the windows have no glass so he could live as much outdoors as indoors and birds could fly in and out as they pleased.

He named the place Goldeneye, not after the duck of the same name but after a plan he'd developed for the defense of Gibraltar during the war, and after one of his favorite books, *Reflections in a Golden Eye*, the 1941 novel by Carson McCullers.

It was the perfect spot for a man who loved watching birds, from the kling-klings (Jamaican grackles) that hung out on his patio to the array of hummingbirds that darted back and forth between the hibiscus and bougainvillea Fleming had planted.

Here, from mid-January to mid-March for thirteen winters, Fleming wrote his world-famous 007 novels while on paid leave from his post as foreign news manager for the *Sunday Times* in London. Here, he entertained such celebrities as Graham Greene,

Chapter 001

Kling-klings (Jamaican grackles) have been getting into mischief at Goldeneye for as long as anyone can remember. Fleming featured them in *The Man with the Golden Gun*. Photo by author

Ian Fleming's bulletwood desk at Goldeneye, where he wrote his 007 novels from 1952 to 1964. *Mark Collins photo, courtesy of Island Outpost*

Errol Flynn, Truman Capote, Katharine Hepburn, British prime minister Anthony Eden, and Princess Margaret.

And here, on February 5, 1964, unexpected visitors appeared at Fleming's front entrance.

The timing couldn't have been better, at least as far as Fleming was concerned. On that sunny Wednesday morning, a film crew from the Canadian Broadcasting Corporation was interviewing the man with the golden typewriter. For Fleming, a master of self-promotion, the interview dovetailed nicely with the publicity for his latest 007 projects. *You Only Live Twice*, his eleventh spy novel, was less than two months from publication. *Goldfinger*, the third screen adaptation of his 007 thrillers, would arrive that September.

Fleming's housekeeper, Violet Cummings, greeted the visitors, an American couple in their sixties, on the grass-covered drive that led from the main road to the bungalow. The man, dressed in light slacks and a loud patterned shirt that shouted "tourist," asked if the author was home. Cummings hurried inside and interrupted Fleming's interview, exclaiming that "Mister James Bond" was there to see him. Fleming thought she was joking at first, then agreed to meet his guest. The thriller writer was about to meet the man whose identity he'd secretly stolen off the dust jacket of a book on his shelf twelve years earlier. At the scene of the crime.

When Ian Fleming greeted Jim Bond and his wife, Mary, that February morning in 1964, he wore a short-sleeved black guayabera shirt, matching slacks, and open-toed sandals that he had chosen for the TV interview. According to Mary Wickham Bond in her 1966 memoir, *How 007 Got His Name* ("by Mrs. James Bond"), Fleming's

The Canadian Broadcasting Corporation interviewed Ian Fleming at Goldeneye on the same day that Jim Bond arrived unexpectedly. *CBC Licensing*

brow was "beaded with perspiration, his fair hair damp and curling, British fashion, behind his ears."

Why might a cool customer like Fleming break a sweat, beyond the camera lights? He didn't know if this visitor who called himself James Bond was the genuine article and—if he was—why he had shown up unexpectedly.

As Fleming led the couple past the TV crew's station wagon to his one-story home, he turned and wondered aloud: "You're not coming to threaten me with a libel case, are you?"

Mary Bond laughed. "Of course not. I want to see where the second James Bond was born."

The fifty-five-year-old Fleming brought his guests into the house and a main room filled with film cameras, sound equipment, lights, and a television crew of four.

In *How 007 Got His Name*, Mary described what happened next:

"Mr. Fleming, tickled with the drama of the situation, lived up to his penchant for the unusual. He flung out his arms and exclaimed, 'This is a bonanza for the CBC! I never saw the man before in my life, but here he is, the real James Bond, stepped right into the picture!' He turned to Jim and said, 'This'll sell even more of your books and mine!'"

Chapter 001

Bond and Fleming met just once, at Goldeneye, in 1964. *Photo by Mary Wickham Bond, Free Library of Philadelphia, Rare Book Department*

Encounter at Goldeneye

The film crew followed Fleming and the Bonds outside to the terrace, where they stood at the edge of the cliff and looked down to a perfect sun-drenched beach. The main camera man was hopping about in the shrubbery, quietly filming the three of them. Bond, quite oblivious to what was going on, kept talking to Fleming about who knows what. Fleming told him to talk for the camera, but Bond couldn't have cared less.

"A lot of birds were flying about and one could sympathize with Mr. Fleming's next question," Mary wrote. "'What kind of birds are they?' he asked Jim. After all, couldn't any handsome stranger drop in and pass himself off as James Bond?"

"Cave swallows," Bond replied. "A very common species in the Antilles. Do you see the square shape of the tail? If you look closely you'll see a chestnut rump."

Fleming nodded, suppressing a smile. The birdman was for real.

Jim Bond immediately got something off his chest. As he told Pete Martin of the *Sunday Bulletin Magazine* later that year, "I confessed to Fleming right off when I met him: 'I don't read your books. My wife reads them all but I never do.' I didn't want to fly under false colors. Fleming said quite seriously, 'I don't blame you.'"

When the film crew left for lunch, the Bonds took a swim with Fleming and his wife, Ann, sipped rum punch, and dined on a traditional Jamaican dish made of saltfish and the national fruit, ackee, which when cooked looks like scrambled eggs and tastes like a buttery potato.

Joining the two couples were the Flemings' house guests from London, Hilary Bray and his wife, Ginnie. Bond was in good company. Fleming had appropriated Bray's name for a character in *On Her Majesty's Secret Service*, published the previous year.

As Mary Bond recounted in *How 007 Got His Name*, during lunch, Fleming encouraged her,

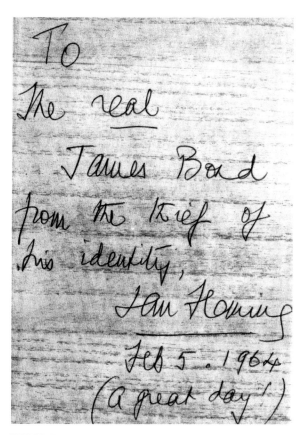

Fleming left no doubt about how he got the name for 007 when he inscribed a first edition of *You Only Live Twice* to "the real James Bond." *Free Library of Philadelphia, Rare Book Department*

a published novelist, to write about her husband's adventures in the Caribbean someday, saying that "whatever he actually did outshines anything I've made *my* James Bond do."

The visit ended a couple of hours later with Bond signing the Goldeneye guest book. Fleming told Bond he deserved a page all his own, adding: "And write it big!"

The American then received a copy of the bestselling author's latest spy novel, inscribed boldly on the fly page: "To the *real* James Bond from the thief of his identity, Ian Fleming, Feb. 5, 1964 (a great day!)."

The Canadian Broadcasting Corporation says the outtakes from that day that include the Bonds no longer exist, but Mary took a snapshot of the two men. They are standing at the front entrance, with two of Fleming's servants waiting in the shadows. Jim Bond stands more than six feet tall, several inches taller than his host, but Fleming has chosen to stand atop the front step, making his guest appear shorter. And although Fleming had suffered a major heart attack three years earlier, he holds a cigarette in his left hand. Both men are smiling. Fleming looks the part of a world-famous author. Bond looks like some guy on vacation who had happened by.

Seeing the contrasts between the bespectacled American ornithologist and the dapper British spy novelist in the photo, it's hard to imagine the two had much in common. Although their paths never crossed again, their lives shared some similarities. Both men were born into ultrawealthy families at the start of the twentieth century. As youngsters, both were intrigued by nature. Both lost a parent when they were young. Both spent their teen years in British boarding schools. (Bond developed a bit of a British accent there—one that stayed with him the rest of his life and led him to pronounce such words as "mobile" like a Londoner instead of a Philly boy.) Both went to college in England and were mediocre students. And after following in their fathers' footsteps, each of them pursued financial careers soon after college.

Perhaps most important of all, both men became authors and found ways to spend their winter months in the West Indies doing what they loved. Fleming created his most lasting work while seated at a typewriter at a red bulletwood desk at Goldeneye. Bond visited dozens of remote islands, equipped only with a pair of eagle eyes and a double-barreled shotgun.

As Morris Cargill recounted in the book *Ian Fleming Introduces Jamaica*, both Fleming and Bond said how much they'd enjoyed that February day. Fleming remarked about the Bonds: "They couldn't have been nicer about my theft of the family name. They said it helped them get through customs."

Hilary Bray wrote to the Bonds: "It was a red-letter day for Ian, as for everyone else. He used to reflect on it happily. And when he got back to England he used to relate it as—Goldeneye, the story of the year, 1964. I don't think anyone considered it his last Goldeneye, though [his days] were numbered."

The encounter in Jamaica has become part of 007 lore. Fleming's twelve spy novels and two short-story collections have sold more than 100 million copies worldwide, but only a few have sold for as much as the book that Fleming inscribed to Bond. A first edition of *Moonraker*, signed by Fleming to Raymond Chandler with the latter's occasional hand-jotted comments in the margins, went for $102,000 at auction in 2004. Four years later, the first edition of *You Only Live Twice* that Fleming inscribed to Bond sold at auction to an anonymous bidder for $82,600 (see chapter 007).

As things turned out, the thriller writer, Goldeneye, and the ornithologist also had second lives, and that inscribed copy has taken on a mysterious life of its own. Fleming died after a heart attack in August 1964. According to biographer Andrew Lycett in *Ian Fleming*, his last words, directed to the ambulance attendants on the way to the hospital, were vintage Fleming: "I'm awfully sorry to trouble you chaps."

But his legacy lives on in the $8 billion 007 empire anchored by the longest-running franchise in movie history. Over the past half century, the Fleming estate has hired five authors to write more than two dozen James Bond thrillers.

In 1976, Fleming's estate sold Goldeneye to Jamaica's Chris Blackwell, the founder of Island Records and the man who introduced the world to Bob Marley and reggae. Blackwell had shown the property to Marley, thinking that the global recording star would want to buy it. When Marley declined, Blackwell bought it himself, changed the name to GoldenEye with a capital *E*, and turned it into one of the most exclusive resorts in the world. While visiting Blackwell at GoldenEye, the pop singer Sting wrote "Every Breath You Take," a song not about love and devotion but an obsessive stalker. Several years later, Bono and the Edge, of the rock band U2, wrote the theme to the seventeenth 007 film there, naming it "Goldeneye."

The property is now a 52-acre destination for the very wealthy. A stay in the Fleming Villa, complete with two satellite cottages, swimming pool, private beach, and tropical garden, starts at more than $5,800 a night. The resort is a few kilometers from two other landmarks: James Bond Beach and Ian Fleming International Airport.

As for the ornithologist whose name became more famous than he was, "the Real James Bond" died nearly a quarter century after Fleming, at age eighty-nine. Over those final twenty-five years, the proper Philadelphian increasingly regretted the 007 connection, especially after *Goldfinger* and *Pussy Galore* landed in American movie theaters and secret agent Bond went from passing fancy to cultural icon.

This portrait of a young Jim Bond was painted by his uncle, noted artist Carroll Sargent Tyson Jr. *Courtesy of David Contosta*

CHAPTER 002

The Bonds of Philadelphia

Blue-eyed, brown-haired James Bond was born into money at the dawn of the twentieth century. Unlike a certain spy of the same name, he went by "Jimmy" or "Jim" his entire life. The twentieth century would be known as the American century, shaped by two cataclysmic wars and incredible technological advances—from automobiles and airplanes to high-powered computers and a ubiquitous internet—and Bond would live through nearly nine decades of it.

The Bond family fortune was enormous, but for the youngster it was ultimately far less significant than its location—Philadelphia, a city long known as the cradle of American ornithology. Thanks to such pioneer ornithologists as author-illustrator Alexander Wilson, the legendary John James Audubon, John Cassin, and museum founder Thomas Say, Bond grew up in an environment where birds and natural history were revered.

Jim Bond was born into a long line of Bonds that dates from as early as the 1600s. His ancestors are said to include Thomas and Phineas Bond, brothers who played key roles in founding the American Philosophical Society, the University of Pennsylvania, and Pennsylvania Hospital. (Bond once told David Contosta, author of *The Private Life of James Bond*, that "Phineas Bond was the most important member of our family.")

The family tree branched into Uruguay in the 1800s, when, according to Contosta's biography, at least two Bond brothers born on the East Coast lived in Montevideo. One was Joshua Bond, the American consul there. The other was Dr. James Bond (Jim Bond's great-grandfather), whose son Francis (Jim Bond's grandfather) also became a doctor and married Sarita Josefa McCall, daughter of a Spanish diplomat serving in Mexico. Francis and Sarita lived in Uruguay and had three children, the youngest of whom was Jim Bond's father, also named Francis, who moved to Philadelphia in the 1880s as a teenager.

Young Francis (Bond's father) quickly worked his way up the social and financial ladder. In 1892, at age twenty-six, he became one of three founding partners of Edward B. Smith & Company, with a seat on the local stock exchange and connections to the wealthiest Philadelphians. Decades later, during the Great Depression, the company merged with Charles D. Barney & Co. to become Smith Barney, one of the largest investment banks in the world.

Francis Bond cemented his social and financial position in 1896 with his marriage to Margaret R. Tyson, cousin of artist John Singer Sargent and granddaughter of John A. Roebling, designer and builder of the Brooklyn Bridge. The wedding, in the heyday of the Gilded Age, was big news.

Jim Bond told author David Contosta that he was a descendant of notable early Philadelphians Phineas and Thomas Bond (*above*). Portrait by Kevin P. Lewellen, courtesy of the Thomas Bond House in Philadelphia

Francis E. Bond, Jim's father, was a partner at E. B. Smith, which later became Smith Barney. *Courtesy of Gwynedd Mercy University*

The High-Falutin' Phillies

In March 1903, Francis E. Bond joined a group of millionaires who purchased the Philadelphia Phillies. A Pittsburgh newspaper reporting on the sale described Bond as "cotillion leader, beau and arbiter of the exclusive assembly set that rules Philadelphia society." Other owners of the Philadelphia Base Ball Club and Exhibition Company included W. Lyman Biddle ("He's a Biddle, and hereabouts that's enough"), Robert K. Cassatt ("son of the great president of the great Pennsylvania Railroad"), and James W. Pau Jr. ("of the house of Drexel").

The Baker Bowl in Philadelphia was the scene of a tragic bleacher collapse in 1903, when Bond's father was an owner of the team. *Library of Congress, Prints & Photographs Division [reproduction number LC-DIG-ggbain-20087]*

The wealthy new owners were soon mocked by an anonymous writer for the *Washington Times,* who wrote: "Should the venture prove successful—and there is no reason why it should not—the future game will be reported in newspapers something like this:

> One of the most fashionable functions of the season was yesterday's game at the Philadelphia Ball Park. It was the occasion of the debut of several charming West Walnut Street belles, and the toilettes observed in the pavilion were exquisite, foulards and tulles predominating.
>
> As cards for the event had been issued under the supervision of the patronesses, the crush was not too great. Tea and bonbons were served after the fifth inning, and the affair was altogether delightful.
>
> The game itself was very interesting. When the home team came upon the field, clad in their silk hats, frock coats and patent leather shoes, there was a ripple of applause from daintily-gloved hands.
>
> The opposing team, which was from Chicago, met with some criticism, however, from the fact that several of its members wore evening clothes, with opera hats and russet shoes.
>
> The umpire was attired in white flannels, canvas shoes and a yachting cap, so that the entire picture was harmonious and pleasing to the eye. . . .
>
> The game was called in the eighth inning, as there were a number of dinner parties for the evening."

Chapter 002

The real Phillies fared worse that year, finishing seventh with a record of forty-nine wins and eighty-six losses. Their season was also marred by tragedy. A jerry-built balcony that was part of the third-base stands at the Huntingdon Street Baseball Grounds (later known as the Baker Bowl) collapsed on August 8 during the second game of a twin bill against the Boston Beaneaters. A dozen people died, and more than 200 were injured. Fearing lawsuits, the ownership—including Bond's father—sold the team.

The disaster, which became known as Black Saturday, was a major influence on the design and construction of twentieth-century ballparks, which relied on reinforced concrete.

The Bonds' four-story brick townhouse at 1821 Pine Street, a few blocks from tony Rittenhouse Square, was the scene of a raft of fancy events during the social season. Francis and Margaret had three children: Margaret (Maggie), born in 1897, Francis Jr., born in 1898, and James, born on January 4, 1900.

Summers were spent on Mt. Desert Island in Maine. So many wealthy Quaker City residents summered and partied there that it was nicknamed Philadelphia on the Rocks.

To keep track of the Bonds and their wealthy friends, one only had to read the

Young Francis Jr. and Jim dressed in seashore attire. *Courtesy of David Contosta*

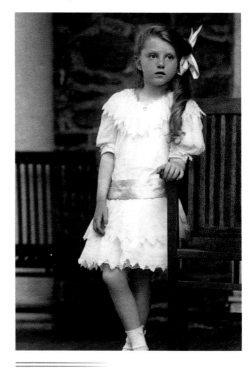

Maggie Bond died on a family Maine vacation in 1904. *Courtesy of David Contosta*

high-society section of the local papers, where they were mentioned in connection with equestrian events, dog shows, fancy balls known as assemblies, carriage parades, and a variety of other charity events.

In early September 1904, the Bonds' utopia came crashing down. While Margaret Bond and the three children were vacationing in Maine, Jimmy's older sister, Maggie, fell deathly ill from a ruptured appendix. The headline in the Thursday, September 8, *Inquirer* tells the story: *Father Raced to Dying Daughter; Philadelphia Banker Traveled Special Train and Yacht to Maine in Time to See Her.* The article concluded: "The run from Philadelphia was made in 10 hours. Mr. Bond found the little girl still alive, but the physicians had given up all hope. She died yesterday morning."

The family never recovered. At first they reacted to the tragedy by building a bigger country mansion in Lower Gwynedd Township in Montgomery County. The enormous house was designed by distinguished Gilded Age architect Horace Trumbauer, whose commissions later included the Philadelphia Museum of Art and the flagship library at Harvard University. While their new house was under construction, the family spent the spring and autumn at a nearby country estate called Spring House Farm. The Bonds' new Willow Brook estate was so noteworthy that it merited a fawning, seven-page spread in the October 1909 issue of *American Homes and Gardens* magazine. The property boasted stables and formal gardens, as well as hundreds of acres of woods and fields where Jim explored and collected butterflies and bird's eggs.

It's unclear when Francis Sr. began to hit the bottle and to womanize, but he retired as a special partner of Edward B. Smith & Company in December 1909 and resigned from the boards of several other companies. A brief article in the *Inquirer* about the "retirement"—Bond was forty-six—noted cryptically that "Mr. Bond will go abroad for a year or two."

That winter, the Bonds made an extended trip to Europe, after which Margaret Bond and her two boys spent the summer in Maine. In early 1911, Francis trekked to Venezuela's Orinoco delta on a three-month expedition to collect birds and other animals for Philadelphia's Academy of Natural Sciences. He did most of the hunting, and assistants took over the bird-skinning and chores.

As Robert McCracken Peck and Patricia Tyson Stroud noted in *A Glorious Enterprise*, the story of the academy, "Although he was not close to his father, the young James Bond could not have helped but be impressed by the acclaim given his father's expedition (not to mention the live howler monkeys he brought back home . . .)." It was about then that doctors diagnosed Margaret with cancer. Within the year, she was dead at age forty-one.

Although she was celebrated in her wedding notice in the *Inquirer* sixteen years earlier as "not only a girl of remarkable beauty but a great favorite in society," Margaret Bond received only a brief death notice in the April 29, 1912, *Philadelphia Inquirer*:

"BOND—Suddenly, 27 inst [this month] at Gwynedd, Pa, Margaret R. Tyson, wife of Francis Edward Bond daughter of Carroll S. Tyson. Services at Church of the Messiah, Mon 4 p m."

Young Jim Bond had been attending the Delancey School, one of the elite schools where proper Philadelphians living in the Rittenhouse Square area sent their sons. That

Name.		Clubs.	Form.	Address.	Entered.
	THE SCHOOL.				9
Bond, James		O. H., H.	II	London, Eng.	1912

This volume of *The Record* of St. Paul's School noted that Bond's father and new wife had moved the family to England. *Courtesy of St. Paul's School*

autumn, he and his older brother were shipped off to the exclusive, boys-only St. Paul's School just outside Concord, New Hampshire. The picturesque, 2,000-acre campus filled with classic brick buildings was 68 miles north of Boston and more than 350 miles from twelve-year-old Jim's home.

In his book *Philadelphia Gentlemen*, fellow Philadelphian E. Digby Baltzell, who matriculated in 1935, characterized St. Paul's as the "oldest, largest and wealthiest of the Episcopal boarding schools," "in the tradition of Harrow." A noted sociologist, Baltzell may be best known for popularizing the acronym that would become synonymous with upper-crust Philadelphians—WASPs (White Anglo-Saxon Protestants).

For Jim, the school likely offered at least one major attraction. "St. Paul's School was and is a wonderful place to bird, and it may have nurtured Bond's interests as it perhaps did later naturalist graduates, including S. Dillon Ripley and John Hay," says St. Paul's alumnus Harry Armistead (more about Ripley in chapter 008, "Twitchers & Spooks").

Noted ornithologist Francis Beach White—who taught at St. Paul's from 1896 to 1942, kept taxidermied birds in his study and curated the school's collection of birds from as far away as Central America and South America—was likely the perfect mentor for the young birder. A 1948 article in St. Paul's alumni magazine noted that "those who did not have athletic practice that afternoon would perhaps go out to White's 'hut' in the woods for an afternoon of observing and banding birds, or for a drive through the countryside in the old Model-T which took ruts and cross-country trails in its stride."

Meanwhile, back in Philadelphia, Bond's widowed father became smitten with Florence Eeles, a British widow with two children of her own. He proposed marriage and raised many Quaker City eyebrows by offering the Willow Brook estate as a wedding present. She accepted both.

The couple married in 1913 under the towering spire of St. Mary Abbots Church in Kensington, London. Florence insisted the Bonds move to England, uprooting Jim and his older brother Francis Jr. at a time when the winds of WWI billowed across Europe. A shy and lonely Jim Bond was sent off to Harrow School in North East London, where he was teased about his accent.

As David Contosta wrote in his 1993 biography, *The Private Life of James Bond*, "The English boys mocked his American accent, all the while insisting that America was a savage and uncouth land filled with wild Indians and the dregs of European society. The worst of the teasing stopped only after Jim became so enraged that he

Bond's carved name can still be seen on a wall at Harrow. *Photo by Mark Ridgway*

Young Bond moved to England and attended Harrow after his father remarried. *Courtesy of Harrow School Archive*

grabbed a pen knife and stabbed one of his tormentors in the arm. From then on, most of the boys respected him for standing up and fighting back."

There are few traces of Jim Bond's time at Harrow. He is mentioned in the school newspaper, *The Harrovian*, at least once, for playing on a sports team. In Druries, the house where Bond lodged, his name is carved into a board on the wall in Call Over Hall, where students assembled three times a day for roll call and notices. Druries housemaster Mark Ridgway rates Bond as a "most illustrious old boy," along with "Lord Byron, Viscount Palmerston, Robert Peel, John Profumo, Lord Butler and a few notable cricketers."

For most of Bond's time at the elite school, the nation was engulfed in WWI, which claimed the lives of almost a million Britons—more than any other conflict in history. As the war raged on the mainland, it hung like a pall over Harrow. At morning assemblies the headmaster would read the names of former students on the ever-increasing lists of casualties. Bond belonged to a military training program, which put him on a track to become a junior officer. According to the Harrow School Register, Bond left in 1918 and "joined Army temporarily."

Next, Bond enrolled at Cambridge University's Trinity College, where he studied economics and honed his marksmanship skills as a member of a hunting club.

According to Cambridge archivist Jonathan Smith, Bond was "admitted as a Pensioner [a fee-paying student] on 8 January 1919 on the tutorial side of Gaillard Lapsley. He does not seem to have read for Honours, but for the Ordinary Degree, getting straight third classes [in examinations in history and political economy]. He graduated BA in

1922." The archivist describes "straight third classes" as the lowest class (grade) above a fail, so Bond had done just about enough to get by.

Bond kept a hunting dog at Trinity and was the only American invited to join the Pitt Club, a small and exclusive dining club and hunting group. Future Pitt Club members included two Brits—Guy Burgess and Anthony Blunt—who went on to great infamy as part of the notorious Cambridge spy ring.

After graduating from Cambridge with a degree in economics, Jim Bond returned to his native Philadelphia in 1922 and followed his father's career path, taking a job in the Foreign Exchange Department of the Pennsylvania Company, a major investment bank and the oldest bank in the United States. Bond was tall, thin, and handsome, with an accent that was "an amalgam of New England, British, and upper-class Philadelphia," according to ornithologist Kenneth Parkes's 1989 obituary of Bond.

Harry Armistead, who lived near Bond in the 1980s in Philadelphia, describes Bond's accent as "Long Valley lockjaw—much in the same manner as the character of the millionaire spoke on the old *Gilligan's Island* TV show."

In March 1923, Bond's father, Francis, died in England at age fifty-eight, perhaps the result of all those years of heavy drinking. Jim, the bearer of the bad family tidings, immediately notified the *Philadelphia Inquirer*. Years earlier, the newspaper had trumpeted Francis Bond's every cotillion, horse show, and financial transaction. Now, the senior Bond received an obituary two paragraphs long, far longer than his first wife's, but still a footnote. An *Inquirer* columnist wrote a more fitting epitaph in "Girard's Talk of the Day" a week later: "Few men so widely known in Philadelphia only a dozen years ago dropped so completely out of it as did Francis E. Bond."

Jim Bond at age nineteen, just before entering Cambridge University. *Courtesy of David Contosta*

Soon after, Jim Bond learned his father had left his entire fortune to his second wife. He was on his own, with a boring job and scant folding money in his pocket.

Bond biographer David Contosta summed up the situation: "His parents had danced at the assemblies and had been accepted into the city's highest social circles. Jim would enjoy the same unquestioned social acceptance by Philadelphia's social elite, a condition that he took for granted and that helped give him an air of nonchalance and social ease that few men and women ever experience. These were accumulated advantages that would assist Jim throughout his life. Yet his father's misbehavior, his mother's untimely death, and his own involuntary removal to England doubtlessly contributed to a shy and introspective nature."

In short, the boy who had been an ultimate insider had become an outsider. Bond was an odd duck, to be sure, and soon to be an adventurous one.

A Remarkable Uncle

Bond's uncle Carroll Sargent Tyson Jr. was a noted artist and collector. *Courtesy of David Contosta*

The largest influence on young Jim Bond was likely his uncle, artist Carroll Sargent Tyson Jr. (1878–1956). Tyson studied art at the Pennsylvania Academy of Fine Arts under William Merritt Chase, and then in Europe, where his friends included Mary Cassatt and Claude Monet. He began to collect their works as well as other impressionist paintings.

Tyson's cousin was the renowned portraitist John Singer Sargent, and Tyson tried his hand at portraits as well. In 1912, he painted Helen Roebling, granddaughter of the builder of the Brooklyn Bridge. Artist and subject fell in love and got married.

Since childhood, Tyson had spent his summers on Mt. Desert Island in Maine. After the death of Jim Bond's mother (Tyson's sister), Jim would join his aunt and uncle at Birchcroft, their mansion in Northeast Harbor, and Tyson's well-appointed hunting lodge in nearby Southwest Harbor. Like his nephew, Tyson was fascinated by ornithology. In 1918, perhaps inspired by Audubon's famous *The Birds of North America*, Tyson began painting watercolors of birds, both from nature and from stuffed specimens, choosing particularly those associated with his Maine environs, both native and migratory. Grandson Charles Tyson Jr. shared this family story:

"One summer afternoon at my grandparents' house in Northeast Harbor, Grandmother and three of her friends were playing bridge on the covered porch when my grandfather barged through the bridge game in his usual disreputable painting garb. In a fit of pique, my grandmother berated him and said, 'By the way, I'm sick and tired of those messy oils and the landscapes you insist on painting. Won't you please do something else?'

Tyson's acclaimed set of chromolithographs of the *Birds of Maine* included this portrait of a snowy owl. *Courtesy of the Philadelphia Museum of Art, Gift of Pauline Townsend Pease, 1979-130-1n. © Estate of Carroll Sargent Tyson*

"The next day, she went out and bought him a set of the most expensive watercolor paints, paper, and brushes she could find, and presented them to him. Thereupon, my grandfather called Jim Bond, and the two of them cooked up the idea of an elephant folio of birds of Mt. Desert Island. Jim provided over a hundred species of birds, trapped and stuffed them, and he and my grandfather posed them in natural settings. My grandfather then made, I believe, 108 sketches, some of which were only partially finished and many were fully painted. They then selected twenty of the paintings they felt were the best. These were sent to an engraving and printing firm [Otto Hoelsch] in Milan, Italy, where plates and prints were made, eventually ending up in the portfolio. And all because of an altercation at a bridge game."

The portfolio of twenty chromolithographs, *The Birds of Mt. Desert Island, Acadia National Park, Maine*, was published in a limited edition of 250 copies, after which the plates were destroyed. On Mt. Desert Island, the Wendell Gilley Museum in Southwest Harbor and the Northeast Harbor Library have complete framed sets of Carroll Tyson's bird prints on display. Northeast Harbor summer resident T. Garrison Morfit (old-time TV personality Garry Moore) came across the portfolio in Bermuda, and he and his wife donated them to the library in Northeast Harbor. Summer resident Martha Stewart also has a complete framed set of the chromolithographs, which she describes as "extraordinary and beautiful." Small wonder that Tyson is considered the John James Audubon of Maine, and that some of his Mt. Desert Island prints sell for thousands of dollars. Although one of Tyson's paintings, *Hall's Quarry*, is in the White House Collection, he is better known for a work that he owned—Vincent Van Gogh's *Fourteen Sunflowers*, one of the seven versions of sunflower still lifes that the Dutch post-impressionist painted in 1888 and 1889. Tyson paid the princely sum of $45,000 for it on a trip to Europe in 1928—on the same day he also bought major paintings by Paul Cézanne, Édouard Manet, and Berthe Morisot.

In Martin Bailey's *The Sunflowers Are Mine*, Tyson's son-in-law, Louis Madeira IV (Bond's cousin by marriage), recalled that Tyson hung the Van Gogh behind his chair in the dining room so he wouldn't have to look at it while he ate. Madeira said that his father-in-law "thought the painting crude and untutored." There is still speculation as to whether the Van Gogh was a tad too modern for Tyson's tastes, or the story was a family joke. Many works from the Mr. and Mrs. Carroll S. Tyson Jr. Collection are on view in the Philadelphia Museum of Art's Resnick Rotunda, including *Sunflowers*, Renoir's *The Large Bathers*, and Monet's *The Japanese Footbridge and the Water Lily Pool, Giverny*. Bond and Tyson also collaborated on the eighty-two-page *Birds of Mt. Desert Island, Acadia National Park, Maine* (see chapter 005).

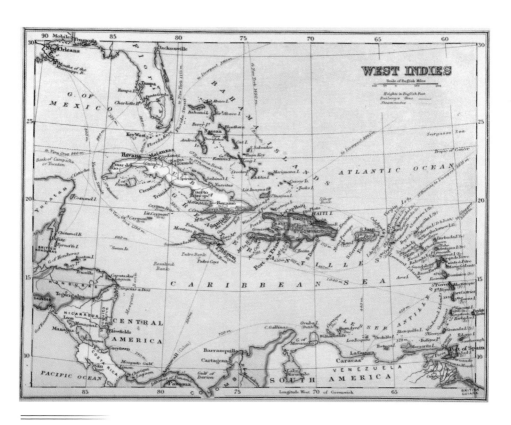

A map of the West Indies at the turn of the twentieth century. *Courtesy of the British Library*

CHAPTER 003

Have Gun, Will Travel

In 1925, Jim Bond made a life-changing decision to quit his banking job so he could explore the Caribbean. Although Bond's father had left his entire fortune to his second wife, both sides of Jim Bond's family tree sprouted money, and perhaps it was inevitable that he inherited some from one of his late mother's relatives.

By shrewdly investing the $25,000 nest egg, he generated enough income to buy his freedom. He accompanied an aristocratic pal named Rodolphe Meyer De Schauensee on an expedition to the lower Amazon jungle in Brazil to collect bird skins for the Academy of Natural Sciences of Philadelphia. The adventure, similar to one that Bond's father had undertaken for the academy fourteen years earlier, also targeted reptiles, fish, insects, and such rare mammals as black jaguars and bush dogs. To pay for the trip, they planned to collect live specimens to sell to zoos, including a strange, almost prehistoric bird called a hoatzin.

The expedition, via a 100-meter-long, steam-powered cargo ship, turned out to be perilous. At one point, a legendarily lethal fer-de-lance popped up on the trail just inches away from Bond's feet, and he had to shoot the 5-foot-long snake before it could strike. Days later, while preparing to bathe in a jungle stream, Bond saw a giant boa constrictor gliding out of the understory. His reaction, as he recounted for the *Philadelphia Daily News*: "I turned and ran. Later I found out the boa was as scared as I was. It darted off in the opposite direction."

The two budding explorers managed to return with many valuable specimens, including a live, 23-foot-long, dark-spotted anaconda and two live hyacinth macaws, but no hoatzin or bush dog. The anaconda turned out to be a new species that would be named in De Schauensee's honor, *Eunectes deschauenseei*. Bond and De Schauensee sold the live specimens to zoos both public and private. One such customer was a New Jersey–based coffee manufacturer named George Washington, who invented an early incarnation of instant coffee and spent part of that fortune keeping exotic animals, including a cheetah that was said to eat breakfast with him at his fenced-in estate.

Bond and De Schauensee broke even and talked of further adventures. Both soon latched on to the academy's ornithology department. Their boss was the renowned Witmer Stone. For the next half century, the pair worked in the academy's stately brick building at the corner of North 19th Street and Race Street.

De Schauensee, whose fascination for birds included keeping an aviary of tropical species at his mother's house, would serve as curator of birds for most of that time. Bond, lacking his colleague's deep pockets, decided to focus on the West Indies, a group of more than one hundred islands scattered across the Caribbean, including Cuba, Jamaica, the Bahamas, the Dominican Republic, and Haiti.

A key factor in Bond's decision was his discovery that a Royal Mail Line steamship between New York and Trinidad stopped twice a month at many of the larger islands. For $125, he could buy an open-ended round-trip ticket and stay as long as he liked on any island he wanted to explore.

On Bond's 1925 US passport, his complexion was listed as swarthy. Similarly, on his 1942 draft card, he checked "white" for race and "dark" for complexion. *Courtesy of David Contosta*

Like his finances, Bond's love life apparently hadn't amounted to much. By one account, Bond fell for an attractive young woman from a wealthy Philadelphia family after his return from Cambridge, but marriage was not in the cards. Bond thought he lacked the means to support her in the accustomed manner, and a family would have tied him down when all he wanted to do, it seemed, was visit the Caribbean and document its bird life.

Bond devoted the next decade to studying the birds there, using ornithologist and bird collector Charles B. Cory's research from the 1880s as a starting point. The goal was ambitious, given the number of birds (some 300 species), the number of islands (beginning with twenty-eight larger ones), the remoteness of so many of them (the West Indies spans 2,000 miles of water), and the fact that Bond tended to get violently ill for the first couple of days aboard a boat.

Beginning in 1926, Bond took more than one hundred expeditions to the West Indies via steamship, despite his seasickness. He traveled so often that one of his superiors at the

Have Gun, Will Travel

On Bond's first expedition, to South America, he sought a distinctive bird called a hoatzin. *Photo by Kevin Watson*

The Academy of Natural Sciences of Philadelphia was Bond's home base for half a century. *Public Domain*

academy complained in an August 1931 letter to Stone that "I find it difficult to keep up with him, as he never tells me anything of his plans."

In *Proceedings of the Academy of Natural Sciences of Philadelphia* (vol. 150, 2000), Bond reflected: "There were far too many Caribbean trips to itemize. I went somewhere in the West Indies two or three times a year, but I could not tell you now where or when."

Hunting birds on foreign islands required permit paperwork—often a difficult proposition without the benefit of reliable phone service. A sample of the paperwork: the Free Library of Philadelphia's archives include a medical certificate signed by a ship's physician on the Dutch steamship *Oranje Nassau*, traveling from Haiti to the Bahamas. He examined Bond in August 1929 and attested that "the said Mr. Bond has not any loathsome, or dangerous, or contagious, or infectious diseases."

Bond's travels and adventures were old-school extraordinary, as he later recounted to his wife, Mary Wickham Bond, whom he married in 1953. In her 1971 memoir *Far Afield in the Caribbean*, she wrote that after arriving in the West Indies, Bond would spend up to nine months at a time exploring islands that ranged from enormous Cuba to tiny Utila, an island off Venezuela. He traveled between islands on tramp steamers, canoes, and primitive sailboats. He made so many trips and stopped at so many ports that once when he stepped onboard the ship in Dominica, the purser said: "You again?"

Unlike his father's expedition to Venezuela, which included a botanist and a bird collector (who skinned and prepared the more than 500 birds collected on the three-month trip), Bond mostly fended for himself. Like most working-class field biologists of his day, he invariably roughed it, drinking water from streams, slogging through crocodile-infested swamps, climbing mountains on foot or horseback—and swatting mosquitoes along the way. He wore "his usual twill trousers, long-sleeved cotton shirt, and no hat," Mary wrote, and seldom carried camera or binoculars—just his double-barreled shotgun.

On his 1928 expedition, Bond found a mentor in Erik Ekman, a botanist from Stockholm who had dedicated himself to studying the plants of Haiti and the Dominican Republic. Despite bouts with malaria, pneumonia, and blackwater fever, Ekman added 1,000 plants to the known flora of the island before his death in 1931 at age forty-six. More than eighty-five years later, a street in Santo Domingo is still named for him: the 2-kilometer-long *Calle Erik Leonard Ekman*.

Bond met Ekman in Port-au-Prince, where both naturalists were staying, and joined the Swede on a hike into the mountains. When the pair reached a stream, Ekman drank the water with gusto while Bond looked on skeptically.

"This was the beginning of Jim's basic training," Mary wrote in *Far Afield in the Caribbean*. "It was certainly the beginning of a long friendship. Ekman not only taught him where it was safe to drink water but how to get along with island people. He taught Jim how to accept their hospitality and pay board with delicacy, leaving behind only enough coins to pay for what he'd eaten but never overdoing it."

Helping Bond assimilate was the fact that he tended to tan deeply and could pass for a Portuguese, Brazilian, or Spaniard—the result of his Uruguayan paternal grandmother's Spanish heritage.

Other invaluable Ekman advice: never look down your nose at any food or lodging that an islander offers, no matter how unappealing, and always rely on local knowledge. "A good *maître-bois*—master of the woods or guide—knows not only the trails but also the local names of birds, plants and animals," Mary Bond wrote. "These names, handed down by word of mouth for generations, are surprisingly reliable and of the utmost help to the collector."

Bond decided a few things for himself—including the belief that it was more important to study the distribution, behavior, song, and nests of the native birds than to collect as many of them as possible for science. As a result, he shot relatively few birds even though he usually carried a shotgun on his expeditions.

Bond's friendship with Ekman was cemented when the Swede, who suffered from malaria and other ailments, collapsed in the Massif de la Hotte, a large mountain range in southwestern Haiti, while the two were collecting. Bond helped him reach a main road on foot, where they caught a bus to Port-au-Prince 50 miles away. It took ten days of bed rest for Ekman to get back on his feet.

The next time Bond visited Haiti, many months later, he set out to find the rare La Selle thrush, named for Massif de la Selle, an 8,000-foot-high plateau running along the southeastern part of the island nation. At first glance, the bird resembles the similar-sized American robin with its brick-red chest, but this endangered thrush has a spectacular orange eye ring and a more put-together appearance. Think of the American robin as a rough draft of the La Selle thrush.

To search for this dynamic bird, Bond typically set off from the mountain village of Kenscoff, about 15 miles from Port-au-Prince, at dawn. He rode a horse for five hours and then climbed the rest of the way up the mountain on foot. "In my thirties, I could climb without too much trouble," Bond recalled in *Far Afield in the Caribbean*. "In my forties it took twice as long, and in my fifties I didn't even try."

In fact, Massif de la Selle was almost the death of Bond. On the mountain during a six-month stay, he became so ill the locals had to lead him to lower elevations, using a stick to steady him. At one point, the 6-foot-2 Bond weighed just 126 pounds, and he heard one of the porters whisper, "*Le ap mouri*"—local slang for "he will die." His friend Ekman heard news of Bond's sorry condition and had the villagers kill a chicken, boil it, and feed Bond broth in hopes of getting food into his system. The next morning, Ekman draped Bond over a horse and sent him on the 15-mile journey back to Port-au-Prince. Amazingly, Bond reached his hotel. The manager got him to the hospital, where he was diagnosed with jaundice.

Cuba wasn't much more inviting. As part of a nine-month bird-collecting expedition in the West Indies that began in late 1929, Bond visited Cuba's 600-square-mile Zapata Swamp on the southwestern coast by the Bay of Pigs. As Mary recounted in *Far Afield in the Caribbean*, his goal was to collect three species that had just been discovered by Fermín Cervera, a Spanish entomologist there.

A former soldier who had served in Cuba during the Spanish-American War, Cervera had stayed on to become a naturalist. He found that a remote outpost of charcoal producers in La Ciénaga de Zapata was a great place to collect rare insects. Cervera would brave

Bond slept in a charcoal burner's hut in the Zapata Swamp in late 1929, during his search for three bird species that had just been discovered. *Courtesy of David Contosta*

crocodiles and quicksand to venture into the wetlands to set up a lamp, likely fueled by kerosene, to attract moths and other insects at night.

When Cervera checked on the activity at the lamp early one morning, he found a strange wren feeding on the flying insects. He sent the specimen to his friend, herpetologist Thomas Barbour of Harvard University, who realized that it was a new species. Barbour told him to go back for more specimens. In short order, Cervera sent him two more birds previously unknown to science. For birders visiting Cuba, the three birds are legendary and more commonly known as the Zapata wren, the Zapata sparrow, and the Zapata rail.

Bond wanted to see them for himself and try to collect specimens for the academy. By the time he arrived in Havana, Cuba's leader, Gerardo Machado, had gone from fairly elected president to dictator, and a revolt was brewing. Bond was allowed to keep his shotgun but was forced to wait ten days in his hotel before receiving his collecting permit. When the document finally arrived festooned with ribbons, official seals, and fancy signatures, Bond read it carefully. He noted that although the paperwork gave him permission to collect birds and mammals, the fine print stipulated that he needed a separate permit to carry a gun.

Bond was one of the few ornithologists to document the mysterious Zapata rail, now considered critically endangered. *Author's collection*

Bond packed his shotgun anyway and took a bumpy three-hour, 40-mile train ride to the seaport of Batabanó on the south coast, where he paid a fisherman $15 to take him on an overnight motorboat run to La Ciénaga de Zapata.

In the middle of the night, after Bond had fallen asleep in the boat's cabin, a Cuban coastal patrol in a gunboat stopped them and asked Bond why he was there. Doing his best to hide his nervousness, he handed his collecting permit to the officers. They reviewed the elaborate signatures, official seals, and assorted ribbons and gave Bond the go-ahead. Sometimes, bureaucratic flourishes had unexpected advantages.

By morning, the boat reached the mouth of Río Hatiguanico, on the west side of the swamp. They traveled several miles upriver until the men had to switch to a skiff, which took them up a canal to Santo Tomás, a tiny outpost of workers who used makeshift kilns to turn wood into charcoal, a key fuel in those days and still a major export. Bond stayed with the charcoal burners for two weeks, conversing in Spanish and living on coffee and a stew made from hutia, a rodent that can weigh up to 18 pounds. In five days, Bond found all three species.

A newspaper account of the 1931 expedition told of how Bond braved "bottomless bogs in Cuba and a storm that periled his frail canoe in the open sea" before returning with "30 species of birds never before represented in the Academy's collections." Several of the species were from the Zapata Swamp. Bond would later return to Cuba just before the 1961 Bay of Pigs debacle (see chapter 009, "Was Jim Bond a Spy?").

Bond Goes Undercover

Getting a permit to carry a firearm to collect birds on Caribbean islands sometimes took a week, and in one instance, an entire year.

As Mary Bond recounted in *Far Afield in the Caribbean*, Jim Bond prepared for his first visit to Jamaica by completing all the necessary paperwork, along with acquiring letters from the US State Department and the Institute of Jamaica, which oversaw all things cultural and scientific for the British colony. Yet, when Bond arrived in Kingston and tried to obtain a gun permit, the governor complained that Americans never carried engraved visiting cards. The governor then announced he had forbidden all bird collecting because the island was down to its last ten pairs of crested quail-doves, an elusive bird also known as the mountain witch.

As far as Bond could tell, the island's mountain witch population was not threatened, let alone endangered. Jamaica's tourism brochures even advertised dove hunting as a prime activity for vacationers. But sensing the governor would not get off his high horse, Bond gave up and left Jamaica. Without the ability to collect birds in the largest British colony in the Caribbean and the third-largest island overall, the prospects for his ten-year project seemed shaky.

Bond had to pose as a tourist on an early visit to Jamaica.
Courtesy of the Library of Congress; public domain

However, a year later he went undercover, posing as a tourist outfitted in a jaunty pith helmet and shorts. Over one shoulder he carried a camera (a rare occurrence for Bond); over the other, a double-barreled shotgun. When a customs official said he'd have to keep the gun until Bond got the necessary permit, Bond reached into a pocket and produced the latest brochure from the island's tourist board, which promoted such wonderful activities for tourists as fishing, sailing, and dove shooting.

"Where do I get that [permit]?" Bond wondered.

The official pointed to the post office just down the street. He paid ten shillings for the permit, retrieved his shotgun, and headed for the Blue Mountains.

Thus went Bond's travels in the West Indies. With so many peregrinations to so many locales, he encountered many new or rare bird species. On a trip to the small island of Gonâve, just off Haiti in the gulf of the same name, he lived for six weeks with a local tailor and his wife and ate a steady diet of coffee and fish with maize, yams, and tropical beans called *pois congo*, a.k.a. pigeon peas. Bond collected fifteen specimens of a least poorwill, a nocturnal, robin-sized bird in the nightjar family. Before then, scientists had collected only one specimen.

"I followed them by ear," Bond told *Philadelphia Daily News* reporter Jay Apt in a May 1955 article. "I listened to their songs at night and tried to establish where they came from. Then I hunted these areas by day."

Bond's philosophy for reaching the islands was "by any means necessary." On Grand Bahama, seeking to visit several places by boat, including High Rock Settlement, Abaco, and South Andros, Bond bummed a ride on a sloop called the *Rescue*, operated by a wiry American known only as Captain Slim. The bootlegger's modus operandi was to import booze from Great Britain, France, and Spain to Nassau, transfer it to 2-gallon gasoline cans, and then sail on the *Rescue* to meet up on the open seas at night with a rumrunner's speedboat from Florida. He could drop Bond off on the way.

Thus, Bond departed on a boat with a deck piled high with gas cans filled with contraband liquor. The following morning, a dory dropped him off at High Rock, a place Mary Bond described in *To James Bond with Love* as "a desolate place, a mere cluster of shacks on the beach." Bond had a duffel, a hammock, and a gun. One of the local islanders was kind enough to let him stay in his hut, and Bond collected birds with his shotgun on the next five mornings and skinned birds in the afternoon, his usual ritual. Meals usually consisted of huge land crabs caught the night before, then boiled or cooked in a stew. (One Bahamian recipe advises to "scrub with a hard ... brush to remove all dirt.")

The trip to High Rock turned out to be a major success. As Mary recounted in *To James Bond with Love*, Bond collected two specimens of a heretofore undescribed nuthatch in the

Bond reviewing part of the bird collection of the Academy of Natural Sciences. ANSP Archives, Collection 457. *Courtesy of the Academy of Natural Sciences of Drexel University*

scrub-pine forests there. The bird is a close relative of the brown-headed nuthatch of the southeastern United States, and many ornithologists consider it a distinct species. The Bahama nuthatch is the only member of that family found in the West Indies, and now is in such decline that it was feared to be extinct following a 2016 hurricane.

In January 1936, Bond's ten-year plan culminated with the publication of *Birds of the West Indies*. He had taken innumerable cargo ships, banana boats, sloops, and canoes to visit fifty islands scattered across the Caribbean, studied all but four of the 174 taxonomic groups of birds found there, and lived to tell the tale.

Jim Bond on an expedition with his shotgun and local residents in British Honduras (now Belize). *Courtesy of David Contosta*

The Bahama nuthatch, first described by Bond in 1931, is on the brink of extinction. *Courtesy of William K. Hayes*

Bond posing with his double-barreled shotgun. *Free Library of Philadelphia, Rare Book Department*

CHAPTER 004

Shotguns & Arsenic

In a box in the Academy of Natural Sciences Library archives sits a well-used knife owned by Jim Bond. The slim pocketknife, made in central New York State, is a deluxe Premium Stockman Knife #69 made by Camillus Cutlery. The 4-inch knife sports imitation-bone handles and three retractable blades, including a time-worn spey blade inscribed "FOR FLESH ONLY."

Through the centuries, ornithologists have killed birds in order to study them, and until recently the basic tools of the trade hadn't changed much since the 1800s: a double-barreled shotgun to kill the bird, scalpel or scissors or knife to skin it, and arsenic to preserve it.

Here's how Bond worked in the field, as his wife, Mary, described in *Far Afield in the Caribbean*:

"Jim collected in the morning and skinned in the afternoon. He had to work hard to keep up with what he collected, and the icebox took a fiendish delight in thwarting him. There was no shelf and everything had to be laid directly on the ice. When a bottom chunk melted the whole mass shifted, tobogganing the birds, cheese, eggs, and bacon to new positions, some out of reach. This always seemed to happen when we weren't around, except in the middle of the night. Then the most awful crashing noise would make one of us leap out of bed."

Mary wrote that she got upset when arsenic, sawdust, forceps, scissors—and blood, guts, and feathers—were on the kitchen table where they sometimes ate their meals, adding, "The job can't be hurried and the excitement comes when after the skin has been cleaned and dusted with arsenic it is turned inside out. The feather side is pulled over the knobby bone structure of the head like a pullover sweater. Sawdust or cornmeal keep the fingers dry while working and I was astonished at how much manhandling the feathers could stand. When finished the skin is laid on its back with demurely crossed feet, cotton eyes, and folded wings, tagged with its Latin name, the date and place where taken, and the name of the collector."

Bond's pocketknife had a spey blade inscribed "FOR FLESH ONLY." *From the Academy of Natural Sciences of Drexel University archives, ANSP.2011.019; photo by author*

For her readers who found the account a tad gruesome, there was also this: "I stared down at the skins and each one seemed to acquire before my eyes a new personality, different of course from the brilliant creature of air, speed, and song when alive, but of deeper significance after being scientifically classified. Each had achieved a definite sense of enviable immortality."

Humans have been offering birds that "sense of inevitable" immortality ever since the ancient Egyptians mummified falcons, but it wasn't until the sixteenth century that the practice became widespread. That's when French naturalist Pierre Belon devised a recipe for preserving birds—by disemboweling the bird, filling the cavity with salt, and then hanging it to dry before stuffing it with anything from tobacco to peppers. Alas, these bird skins still developed rot and insect damage.

The oldest stuffed bird still in existence is likely an African grey parrot that resides in London's Westminster Abbey. The pet of Frances Teresa Stuart (1647–1702), duchess of Richmond and Lennox, the taxidermied parrot is displayed on a stand next to a wax effigy of the duchess. X-rays have shown that the entire skeleton of the bird, including its skull, remains intact. It is thought to have survived so long because it has been displayed under glass, protected from insects and humidity changes.

In the late 1730s, taxidermy got its first major boost when French pharmacist and naturalist Jean-Baptiste Bécoeur experimented with a wide range of chemicals to determine which were the most effective against insects. He tested fifty chemicals, one on each of fifty specimens, to see if any would remain free of insect damage for four years. Only four of the specimens remained intact. He combined the chemicals—arsenic, camphor, potassium carbonate, and calcium hydroxide—with soap to create a concoction that would eventually be known as arsenic paste.

Carolina parakeet collected by John James Audubon, Fort Leavenworth, Kansas, 1843. Academy of Natural Sciences of Drexel University, Ornithology Department, #136786. *Photo by author*

The use of arsenic as a preservative contributed to the demise of the Academy of Natural Sciences' first main curator, John Cassin, who amassed the world's largest collection of birds for the academy and spent two decades working with the study skins. These were the days before latex gloves and other safety precautions. Cassin died in 1869 at age fifty-five, with arsenic the suspected culprit.

Arsenic had also caused the early death of another noted ornithologist from Philadelphia, John Kirk Townsend, for whom the colorful Townsend's warbler and the drab Townsend's solitaire are named. Townsend died of arsenic poisoning in 1851 at age forty-one, after developing a formula that included arsenic as the "secret" ingredient for preparing taxidermy preparations.

Robert Ridgway, curator of birds at the US National Museum (now the Smithsonian National Museum of Natural History), took a more straightforward approach to both shotguns and arsenic in his twenty-two-page 1891 treatise, *Directions for Collecting Birds*. He advised that the best gun for "all round" collecting is "a 12-gauge, double-barrel, breech-loading shotgun of approved make, with barrels 28 inches long, length of stock and 'drop' to suit the user," and, in case someone wanted to shoot smaller birds, an auxiliary barrel for .32-caliber shells loaded with "wood powder, Grade D, and No. 12 shot." This is the type of shotgun Bond often used in the field.

Ridgway also recommended mixing arsenic with alum to preserve birds, because "poisoning of fingers, through cuts or abrasion of the skin, is far less likely."

Birding the Old-Fashioned Way

For folks in the late 1800s who preferred keeping bird skins in their drawers (so to speak) rather than putting feathers in their hats, a major resource was ornithologist Elliott Coues's 1874 *Field Ornithology*. The subtitle said it all: "comprising a manual of instruction for procuring, preparing and preserving birds."

The heavily bearded Coues (pronounced "*Cows*"), a US Army surgeon for nearly two decades, was a founder of the influential American Ornithologists' Union (AOU) and editor of its quarterly journal, *The Auk*.

The first sentence of *Field Ornithology* takes dead aim at its intended readership: "The double-barreled shotgun is your main reliance." The fourteen-page first chapter is devoted entirely to shotgun use, with subjects ranging from selecting ammo to loading and caring for your gun.

With its gold-embossed owl logo on the cover, Coues's treatise is chock-full of antiquated and cringeworthy nuggets on shooting warblers and "breaking the law unobtrusively." Some highlights:

• On using a contraption called a cane gun, designed to look like a walking stick: "If you are shooting where the law forbids the destruction of small birds—a wise and good law that you may sometimes be inclined to defy—artfully careless handling of the deceitful implement may prevent arrest and fine."

• On marksmanship: "It is finer shooting, I think, to drop a warbler skipping about a tree-top than to stop a quail at full speed."

• On how many birds of the same species you should collect: "*All you can get*—with some reasonable limitations; say fifty or a hundred of any but the most abundant and widely diffused species."

Elliott Coues's classic guide to field ornithology advised, "Arsenic is a good friend of ours; besides preserving our birds, it keeps busybodies and meddlesome folks away from the scene of operations." *Public domain*

- On arsenic: "Arsenic is a good friend of ours; besides preserving our birds, it keeps busybodies and meddlesome folks away from the scene of operations, by raising a wholesome suspicion of the taxidermist's surroundings."

In his book *Biographical Memoir of Elliott Coues*, Joel Asaph Allen, first president of the AOU, called *Field Ornithology* "without doubt, one of the most useful and popular manuals of ornithological field work ever put forth."

A major proponent of the shotgun school of the late 1800s was ornithologist and collector Charles Barney Cory, a Boston Brahmin who wrote the first *Birds of the West Indies* (see next chapter). Cory, the son of a wealthy importer, developed his trigger finger early. At age eleven, he saved his money and secretly bought a pistol that he and a friend used to shoot at birds. Five years later, he began amassing a collection that eventually totaled 19,000 bird skins.

In 1902, at the height of his fame as an ornithologist, Cory was asked to speak at a meeting of the Audubon Society, which had been formed to protect birds from market gunning and the fad of using their feathers to adorn women's hats. As recounted in Mark V. Barrow Jr.'s *A Passion for Birds*, Cory declined the invitation, saying: "I do not protect birds. I kill them."

(In a similar vein, consider this anecdote from *Warrior: The Legend of Colonel Richard Meinertzhagen*, by Peter Hathaway Capstick: The twentieth-century British ornithologist, spy, and pathological liar was once asked at a dinner party if he still shot and collected birds.

When Meinertzhagen acted like he hadn't heard his fellow guest, she pretended to fire a gun and added: "You know—bang, bang!" To which he replied: "No, ma'am. Bang!")

CHICAGO NATURAL HISTORY MUSEUM

This field guide to skinning birds, written by former counterespionage agent Emmet Blake, suggested the use of powdered arsenic and other dangerous chemicals. *Author's collection*

Thanks to the efforts of the Audubon societies and other fledgling groups, the public's views on bird collecting grew more discerning, and the federal government intervened to stop illegal poaching and wanton slaughter. First came the Lacey Act, which prohibited so-called market hunters from selling poached game across state lines, followed by the Migratory Bird Treaty Act of 1918, which made it "unlawful to pursue, hunt, take, capture, kill, possess, sell, purchase, barter, import, export, or transport any migratory bird, or any part, nest, or egg or any such bird." The act did allow the world's major natural-history museums and several top universities to continue to expand their global research and build their collections under strict regulations. Bond's work in the Caribbean was one small facet of this effort.

During WWII, the Smithsonian Institution even tried to recruit servicemen to collect specimens, producing a pocket-size softcover book called *A Field Collector's Manual in Natural History*. Think of it as a citizen-science project for American soldiers and sailors. The introduction to the 1944 publication explains: "Many of the men serving in our armed forces who have been sent to all parts of the world have a keen interest in the animals, plants, rocks, and other objects about them, and, as their duties permit, find recreation in examining them."

Inside the 138-page manual are instructions on mammals, birds (including eggs and skeletons), reptiles, amphibians, fishes, acorn worms, leeches—you name it. The largest section, twenty pages, is on birds. For a preservative, the manual suggested arsenic.

In 1949, Emmet Reid Blake (see chapter 008, "Twitchers & Spooks") of the Chicago Natural History Museum wrote "Preserving Birds for Study," a thirty-eight-page pamphlet with a similar aim: to instruct amateurs on how to "prepare themselves for the task of preserving birds collected for scientific purposes." Blake's list of instruments

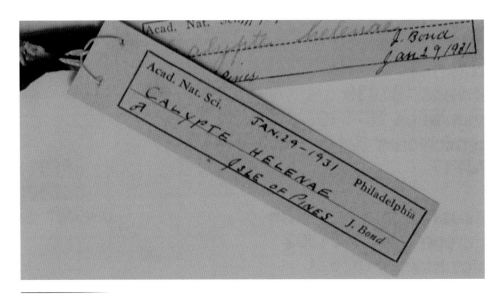

This label is attached to the bee hummingbird that Bond collected for the Academy of Natural Sciences on Cuba's Isle of Pines in 1931. *Photo by author*

and materials alone would have deterred most birders. Suggested chemicals included heavy magnesium oxide, carbon tetrachloride, powdered arsenic, naphthalene flakes, and formalin or alcohol. Blake advises that carbon tetrachloride is the best solvent because it is "extremely safe to handle." Carbon tet, as it was known, has since been banned except for limited industrial applications because of its potential damage to the human heart, blood vessels, liver, and nervous system.

Although Bond was a major proponent of arsenic, he was not as quick to pull the trigger as many of his contemporaries. Mary Bond wrote in *To James Bond with Love* that the "accumulation of study skins was to Jim much less important than acquiring information on the distribution, behavior, song, and nidification [nesting] of the native avifauna of which so little was known. Although usually carrying a gun on his expeditions, he shot few birds." Bond's minimalist approach to collecting had an unforeseen downside, says Nate Rice, the Academy of Natural Sciences' ornithology collection manager:

"To contrast what Bond did, other people were going to places, collecting specimens of several individual species every year, for decades. That's the foundation for our understanding of environmental change, climate change, habitat manipulation, and population genetic changes. In the Caribbean, it would have been very valuable to have, say, one bird of whatever species collected every year on Cuba and Jamaica and the Lesser Antilles over the course of Bond's career."

Those seemingly excessive collections of birds by Charles Cory and others more than a century ago have a silver lining. "If it weren't for guys like that, we wouldn't know a lot of what we do about birds," Rice says. "Those huge collections from the turn of the century and the early twentieth century are a gold mine to us researchers now for genetic studies, for environmental surveys."

As an example of these old specimens' value, Rice cites the study skins John James Audubon prepared for the academy, including specimens of such now-extinct birds as the Carolina parakeet. Researchers are able to sequence the DNA from a snippet of skin on the parakeet's foot to get a genetic profile that lets them see how it relates to other groups of birds. From a snippet of feather they can also determine what its environment was like.

Equally important, Rice says, is that it's impossible to know what techniques will be available to analyze the same study skins in ten or a hundred years. The bee hummingbird that Bond collected in Cuba in 1931—which now sits in a tray of hummingbirds in Cabinet 98 on the climate-controlled fourth floor of the Academy of Natural Sciences—may provide all sorts of new data in 2031 and beyond.

CHAPTER 004

The Carriker Treachery

If every closet has at least one skeleton, then the largest one in Bond's closet belonged to a lean, lifelong ornithologist named Melbourne Armstrong (Meb) Carriker Jr., whose career Bond sabotaged.

Carriker (CARE-ick-er), a colleague of Bond's at ANSP from 1929 to 1938, was the most prolific collector of neotropical birds in history—more than 75,000 specimens. He was also a well-published authority on bird lice, collecting and studying thousands of specimens of these tiny, wingless insects.

Raised in Nebraska, Carriker became interested in birds early and was collecting and skinning them while still in high school. After graduation from the University of Nebraska in 1902, he joined a six-week expedition to Costa Rica, beginning a career in Latin America that spanned fifty years.

In 1927, Carriker and his wife and family moved to Toms River on the Jersey Shore so that their five bilingual children could attend American schools and Carriker might land a job at the academy in Philadelphia. After working as a carpenter for two years, he joined the academy staff as an assistant curator. Carriker was a newfangled ornithologist—one who wanted to be paid for his professional skills, in contrast to colleagues Rudy De Schauensee and Jim Bond, who had enough money to work for free and were thus more likely to keep their jobs when academy money was tight.

Then came the Great Depression. As Bond's ANSP colleague Ruth Patrick told Bond biographer David Contosta, the academy back then "was dirty. It was rundown. Beggars used to come in and sleep on the windows." Patrick said money was so tight that "lights were never left on at night. None. If you wanted to go through the building at night, you felt your way, or if you had a flashlight, that would be fine."

When Carriker's job was in jeopardy, bird curator Witmer Stone intervened, telling the academy's executive director, Charles M. B. Cadwalader, that Carriker's work in Peru was "one of the outstanding pieces of work that our institution has been able to carry out," and that the birds Carriker had collected there represented the finest collection of Peruvian birds in the world.

During a 1934–35 expedition to Bolivia, Carriker sent back more than 2,250 bird specimens—likely more than Bond collected in his entire career. Carriker embarked on a second expedition to Bolivia in 1936. That's when Bond apparently began to play institutional politics, and Carriker's good standing took an abrupt turn. Bond's *Birds of the West Indies* had just been published, and he and De Schauensee were embarking on a new project, *Birds of Bolivia*, which relied principally on the new species Carriker had discovered, as well as the quality and quantity of the bird skins he'd collected.

Rodolphe Meyer De Schauensee and Jim Bond examine a taxidermied unicorn bird collected by Meb Carriker. *Courtesy of David Contosta*

Once Carriker had left for Bolivia and was unable to defend himself, he received a letter from Cadwalader that began as follows:

"Yesterday I was much distressed and disturbed on hearing that some of the birds collected by you in South America were in really bad condition. I give below a list of 10 birds, some of which I have personally examined, and there is no question as to their being in exceedingly bad condition." The reason: not enough arsenic.

"You must use arsenic in the preparation of your specimens and you must use sufficient arsenic to do the job right," Cadwalader wrote. "Otherwise there is no use in continuing this work."

Nate Rice, the academy's ornithology collection manager, believes Bond was the provocateur behind Cadwalader's letter to Carriker. He says the irony is that Carriker was chided for the poor quality of his bird skins, when it was Bond who was terrible at preparing specimens. And if Carriker stinted on the arsenic, he was ahead of his time.

"I think Bond was quite jealous of Meb Carriker. Carriker went all over Latin America for decades and was 'the man' in South America. Carriker submitted a proposal to the academy to start work in the Caribbean, and Bond was totally against it."

If Bond was envious of Carriker, the situation no doubt worsened the following year with Carriker's discovery of a bird that would be bestowed with one of the most enchanted names ever—the unicorn bird. Carriker collected at least four specimens of the turkey-like bird in Bolivia in 1937. Also known as a horned curassow or southern helmeted curassow, it weighed roughly 8 pounds, with wings up to 16 inches long.

Storrs Olson, curator emeritus of the Smithsonian, who in 2007 published the exchange of letters between Cadwalader and Carriker, noted: "No one else could have accomplished what he [Carriker] did, and his reward was getting kicked out of the academy so that Bond and De Schauensee could reap all the rewards."

In October 1939, after the academy had dismissed Carriker, ostensibly for budgetary reasons, Bond and Rudy De Schauensee announced in an academy newsletter that the unicorn bird was a new species, *Pauxi unicornis*. Although the two ornithologists said that the new species had been discovered by Meb Carriker, that fact was lost in the newspaper coverage, as evidenced by this nonbylined wire story in the *Escanaba Daily Press* on October 10 of that year:

> A turkey-like bird with a three-inch horn growing out of its forehead, which was discovered in the jungles of Bolivia, was announced by James Bond and Rodolphe Meyer De Schauensee, curators of birds of the Academy of Natural Sciences of Philadelphia. They call it the unicorn bird: *Pauxi unicornis*, to the scientist.
>
> The bird resembles the new streamlined Thanksgiving bird which was recently developed by the department of agriculture, in its size, which is about eight to 10 pounds in weight. . . . The new species was discovered by a collector for the Academy.
>
> Mr. De Schauensee, in announcing the finding of the unicorn bird, said, 'The mystery surrounding this turkey-like bird is great, particularly in view of the fact that it is edible. Few edible birds escape the natives of a South American jungle. That it should have remained unknown in a relatively well-explored portion of the country is additionally strange.'

Despite the fact that Bond had never been to Bolivia, he and De Schauensee went on to write *The Birds of Bolivia*, published in two parts by the Academy of Natural Sciences during WWII.

Perhaps Mary was remembering this incident a quarter century later when she wrote (in *How 007 Got His Name*) about the day she and Bond visited Ian Fleming at Goldeneye. She observed, "Jim got out of the car, an inch or two taller than 007 perhaps but skinnier and capable of similar ruthlessness if the situation required."

One final irony: The ANSP has a web page dedicated to Charles M. B. Cadwalader's extensive accomplishments, including this feather in his cap: "The largest of the

Cadwalader contributions to the Ornithology Department is the series of specimens collected by Meb Carriker in Bolivia from 1935 to 1938. Nearly 8,000 study skins are included in this spectacular series, all beautifully prepared by Carriker, one of the greatest field ornithologists of all time. Included in the collection are type specimens [the specimen that the scientific name of the bird species is almost always based on] for at least 74 new species of birds found during these expeditions."

Photo by author

CHAPTER 005

Birds of the West Indies

On the shelf above my desk are a dozen books of various shapes and sizes titled *Birds of the West Indies*. Together they weigh 20 pounds and span 12 inches, 4,000 pages, and more than 125 years from end to end.

Two immense volumes serve as bookends and could also serve as doorstops. One is a textbook by ornithologist Charles Cory published in the late 1880s; the other is a large-format photography book by well-known conceptual artist Taryn Simon published in 2013. In between are ten books—seven authored by Jim Bond—whose various incarnations are a study not only in the advances in ornithology, birdwatching, and publishing, but also travel and technology. As far as birdwatchers are concerned, the Bond books, and more recently their successors from Princeton University Press, are the real deal.

Bond's book was printed in eight or more editions (depending who's counting) over nearly six decades, an incredible run. It was finally overtaken in 1998 as the definitive birding book for this region by Princeton University Press's *Birds of the West Indies*, by Herbert Raffaele and a team of four other birding experts and two illustrators. At 512 pages, this identification guide is far more exhaustive than Bond's 1936 original edition or the many that followed.

And the tradition continues. Princeton University Press is publishing an updated version of Raffaele's 2003 book in 2020.

Here are the books, each fascinating, in chronological order by date.

Birds of the West Indies, 1889, by Charles B. Cory, published by Estes & Lauriat. At 328 pages, this hardcover book with two maps and a smattering of black-and-white illustrations was clearly designed for the bookshelf, not the field. All the species are under their Latin names, so that if you were looking for a description of, say, a Cuban tody, you were in for a lot of looking. When his description of the tody refers to "the faint whitish tippings when held to the light," you can be sure he was more accustomed to viewing birds on trays than in trees.

Charles B. Cory wrote the first *Birds of the West Indies* in 1889. *Public domain*

Birds of the West Indies, 1936, by James Bond, published by the Academy of Natural Sciences. Hard cover, 460 pages, with a map on both endpapers showing voyages taken and islands he visited in 1935, plus 159 line drawings by Earl Poole and one color illustration of a Cuban tody.

This is the alpha of Bond's *Birds of the West Indies*, and the first to cover nearly all the nonmigratory birds found in the West Indies. It was as much a classic ornithology text as a field guide. It was published two years after Roger Tory Peterson's classic *A Field Guide to the Birds* and paled by comparison—the dense descriptions and the single color plate of a Cuban tody could not compare with Peterson's handier volume with four-color plates of multiple birds.

Nonetheless, this first edition, complete with iconic white dust jacket, is rare and worth thousands of dollars. By all authoritative accounts, it was this edition that prompted Fleming to name his secret agent for James Bond sixteen years after it was first published.

The first edition came in two bindings. One was bound in gilt-stamped green cloth with small imprint lettering blocked on the spine. The other is gray-green cloth with large imprint lettering and the word "of" stamped on the spine in a large font.

Field Guide to Birds of the West Indies, 1947, by James Bond, published by Macmillan Company. Hard cover, 258 pages, featuring 211 line drawings by Earl L. Poole.

This second edition is far more compact and more in keeping with a modern field guide. On the dust jacket's front flap is a blurb that captures the book's appeal: "Whether you are an ornithologist planning a field trip to the West Indies or a globe trotter anticipating a pleasure trip to these beautiful islands, you will want to tuck this little book into your luggage for quick reference. . . . Its primary purpose is to enable the reader to identify the various birds known to inhabit the Bahamas, the Greater Antilles—including the Cayman and Swan Islands—and the Lesser Antilles."

Birds of the West Indies, various editions from 1960 to 2002, by James Bond, published by Collins and Houghton Mifflin. Hard cover, 256 pages. Featuring eighty full-color illustrations by Don R. Eckelberry on eight plates and 180 line drawings by Earl Poole.

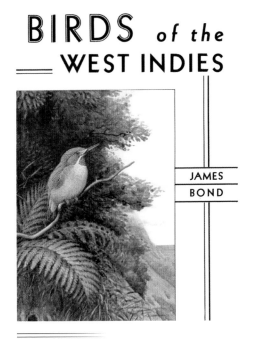 

Front of the dust jacket for Bond's 1936 *Birds of the West Indies*. *Courtesy of Jack Holloway*

Front of the dust jacket for Bond's 1947 *Birds of the West Indies*. *Author's collection*

There was a new publisher for the 1960–1961 edition, and also a British edition, and an overhaul that made it far more practical to use in the field. The two editions have different dust jackets.

The 1961 American edition features a photograph by Mary Bond on the back of the dust jacket. James Bond is wearing a brightly patterned, short-sleeved shirt that looks identical to the one he wore when he and Mary visited Ian Fleming in 1964. What's extraordinary is that Bond is skinning a European cuckoo. It might be the only field guide featuring a photo of its author with a bird that he likely killed.

The later editions were essentially the same, with minor variations, but good luck trying to figure out which edition is which. There was no third edition, for example, but at least three fifth editions, including one in 1995 published by Easton Press that was retitled *Birds of the Caribbean*. Its leather cover has gold lettering and touts its affiliation with the Roger Tory Peterson field guides. On the front cover, James Bond is identified in tiny type—by his last name only.

Birds of the West Indies, 1998, by Herbert Raffaele, James Wiley, Orlando Garrido (Bond's longtime colleague in Cuba), Allan Keith, and Janis Raffaele, published by Princeton University Press and Christopher Helm, London. Hard cover, 512 pages. 86 color plates.

This hardcover identification guide, published sixty-two years after Bond's first *Birds of the West Indies*, is part of a series of old-school Helm Identification Guides and, at almost 3 pounds, a chore to bring into the field.

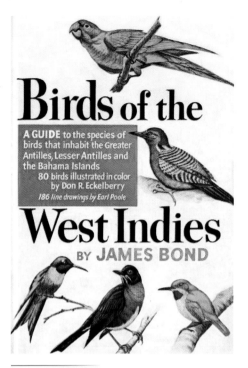

Front of the dust jacket for Bond's 1961 *Birds of the West Indies*. Courtesy of Jack Holloway

Back cover of the 1961 edition featured Bond skinning a European cuckoo. *Free Library of Philadelphia, Rare Book Department*

As with other guides in the series, detailed introductory sections are followed by the plates and information on identification, local names, voice, status, range, habitat, and nesting. A locality checklist provides an at-a-glance, island-by-island guide to the distribution and status of every species.

Raffaele became interested in the birds of the Caribbean when he was eighteen, and later took a trip to Puerto Rico using Bond's book as his field guide. "Before I left, I went to my local library in Jamaica, Queens, and went straight to 598.2 on the nonfiction shelf—I knew all the birding books began with that Dewey decimal number—to look for a book on the birds of the region," Raffaele recalls. "And Bond's book was the book I found. I said, 'Wow, just what I need.'"

Once Raffaele got to Puerto Rico, he was thrilled to find all the birds illustrated in Bond's book, but he had trouble identifying small finch-like birds called grassquits because "all that Bond had illustrated was the glamorous stuff."

Raffaele decided to write his own field guide to Puerto Rico. It was first published in 1983, followed by his *Birds of the West Indies*, the ultimate guide for birding in the Caribbean and a major step in the ornithological literature of this region.

Birds of the West Indies, 2003, by Herbert Raffaele, James Wiley, Allan Keith, Janis Raffaele, and Orlando Garrido, published by Princeton University Press, Princeton and

Oxford. 216 pages. This field guide is a classic example of less is more. Weighing two-thirds less than the previous version, it's filled with useful information and illustrations for identifying birds, including an overview of how common or rare they are, and where they might be seen.

Birds of the West Indies, 2010, written and illustrated by Norman Arlott, published by Princeton University Press. Featuring color illustrations of sixteen birds on the front and two more on the back, this book is billed as an "illustrated checklist," one of a series by Princeton University Press. It's compact and crammed with information, and among its features are eighty color plates featuring more than 550 bird species, a short text that focuses on identifying the bird in the field, and distribution maps for each species.

Its compactness comes at a price—the typeface is tiny, as are many of the illustrations (try to differentiate the Cuban tody from its four cousins among the twelve illustrations on page 97, for example).

Birds of the West Indies, 2013, by Taryn Simon, published by Hatje Cantz, Germany. Two black-and-white Earl Poole illustrations, of a sharp-shinned hawk and a cloud swift that first appeared in the 1936 Bond edition, are featured at the beginning and end of the book.

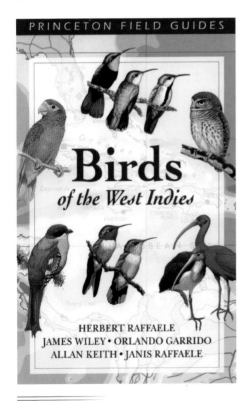

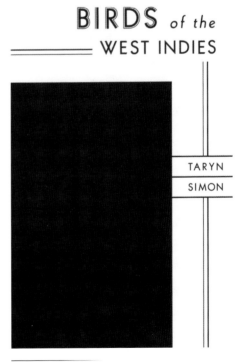

Front cover of Herbert Raffaele's 2003 *Birds of the West Indies*. Author's collection

Front of the dust jacket for Taryn Simon's 2013 *Birds of the West Indies*. Author's collection

A leading conceptual-art photographer, Simon produced a large-format book with a front cover that pays homage to Bond's 1936 dust jacket and reinforces the connection between him and Ian Fleming's 007. A reviewer for *Time* magazine described the first part of the 440-page book as "a meticulous and mesmerizing meditation on materialism, masculinity, and geopolitical movements over the course of the last 50 years, via the vehicles, weapons, and women that have been featured in Bond films during the same period of time."

In the second part of the book, according to Simon's own description, she casts herself as "the ornithologist James Bond, identifying, photographing, and classifying all the birds that appear within the 24 films of the James Bond franchise."

Birds of Mt. Desert Island

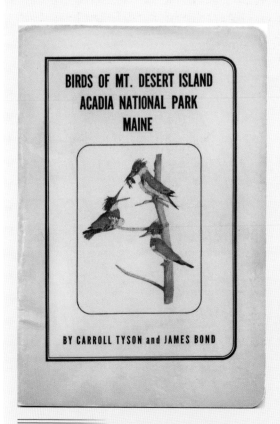

Front cover of Carroll Tyson and James Bond's 1941 book on Mt. Desert Island. *Author's collection*

Bond wrote or cowrote field guides, scientific papers, and checklists for most of his adult life. The most enduring (and endearing) of the lot is an eighty-two-page booklet called *Birds of Mt. Desert Island, Acadia National Park, Maine*, written with his uncle Carroll Sargent Tyson Jr. The paperback, published by the Academy of Natural Sciences in 1941, is not your typical field guide. The authors not only watched the birds—they shot them.

On the very first page, for example, the entry on loons notes: "They dive with great dexterity when shot at, disappearing at the flash of the gun, thus rarely being hit." In a description of the northern goshawk, the authors write: "We have several times disturbed a Goshawk at its feast of grouse and shot the marauder." The common American crow, meanwhile, "is well known and widespread on Mt. Desert Island where it is jokingly referred to as 'soup-meat'!"

Like Bond's original *Birds of the West Indies*, the book is now a collector's item and the progenitor of several subsequent editions, including a 2018 leather-bound reprint of the 1941 first edition. Ralph Long followed suit with his own privately printed version, beginning in the early 1980s. He dedicated the book in part to Bond, "who was the first to inspire my interest in birds and who compiled the early editions of *Native Birds of Mt. Desert Island*."

In the foreword of his forty-two-page paperback, Long wrote that Bond asked him to continue the tradition of "periodically reporting on the status of the breeding birds on Mt. Desert and nearby islands," which "probably surpasses that of any area of comparable size and diversity of avifauna in Maine or any other New England state."

Long's booklet featured artwork by noted nature illustrator Ned Smith and—unlike the Bond and Tyson edition—made no mention of shooting goshawks or using crows for soup meat.

After Long died in the early 1990s, Rich MacDonald, of the Natural History Center in Bar Harbor, continued the tradition of writing about the island's native birds. MacDonald says he's developing the previous booklets into a full-size book of several hundred pages. Not surprisingly, he's a fan of the real James Bond: "When I was about ten years old, I discovered a passion for everything bird," he says. "Not long after that, I found a copy of Bond's *Birds of the West Indies* in a yard sale. Given his name, I just had to learn more about this person."

The only known photo of Bond and Fleming at Goldeneye in 1964. *Photo by Mary Wickham Bond, Free Library of Philadelphia, Rare Book Department*

CHAPTER 006

The Bond-Fleming Bond

The world's most flagrant case of identity theft began in 1952 as a lark, or so the oft-repeated and condensed story goes. As British author Ian Fleming neared his wedding day at Goldeneye in Jamaica, he decided to write a spy novel to calm his nerves. Fleming was writing his first 007 adventure and casting about for a name for his secret agent, when he happened to glance down at his guidebook to Caribbean birds and something clicked.

"On or around 17 February," Matthew Parker writes in *Goldeneye*, his biography of Fleming in Jamaica, "he sat down at his desk in Goldeneye's main room, plucked a name from the author of *Birds of the West Indies*, whose book sat on his shelf, lined up his ream of smart paper and started to write. So began Bond, with the claustrophobic first line of *Casino Royale*."

The name "James Bond" was just the sort of moniker Fleming was looking for, and he never thought to let the real Jim Bond know because Fleming had no idea whether his thriller would ever get published, let alone launch one of the most successful movie and literary franchises in history.

The early 007 novels, first *Casino Royale*, then *Live and Let Die* and *Moonraker*, became bestsellers in Great Britain, but they remained undiscovered in the United States for many years after they were published. A live teleplay of *Casino Royale* aired in October 1954 on the American TV series *Climax!*, with well-known actor Barry Nelson playing an American secret agent named Jimmy Bond and Peter Lorre playing the villain Le Chiffre. In an age of few reruns and no VCRs, DVRs, VHS movies, DVDs, cable movie channels, or Netflix, the fifty-two-minute Hollywood drama disappeared into the ozone as soon as it aired (these days you can watch a YouTube video of the telecast made from an old Kinescope). Similarly, the first American paperback edition of *Casino Royale* was published the following year as *You Asked for It*. As with the teleplay, the book's back cover referred to James Bond as "Jimmy Bond."

Chapter 006

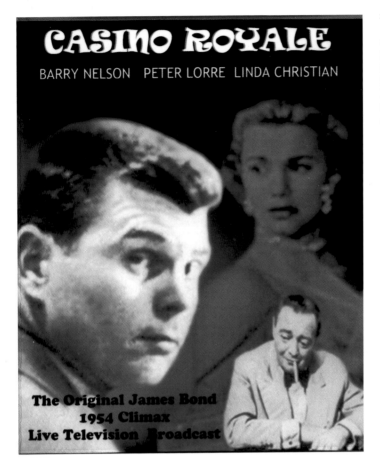

A young Barry Nelson played "Jimmy Bond" in *Casino Royale*, a live teleplay broadcast in 1954 and now available on DVD and YouTube. *Author's collection*

The first hint that Fleming might have based his protagonist on a real person came with the publication of *Dr. No* in 1958. Toward the end of the book, after 007 is apprehended by Dr. No's henchmen, Bond claims he's an ornithologist.

One of his captors replies: "Could you please spell that?"

Jim Bond had no inkling that a game was afoot until the fall of 1960, when a friend sent him a review of the first true field-guide version of *Birds of the West Indies*, from the usually staid *Sunday Times* of London:

> WRONG MAN
>
> "Image" is the new nauseating word. I can barely bring myself to write that James Bond, like practically everyone else mentioned in the newspapers these days, is trying to establish a new image for himself.
>
> To show that his life is not all sadomasochism, Smith and Wessons, and ecrevisse-tails in a white wine and brandy, Bond has revealed himself to be a bird-watcher; and I shouldn't be a bit surprised if his *Birds of the West Indies* which comes from Collins tomorrow, doesn't place him, ornithologically speaking, well above Wolf Mankowitz, but somewhat

below Earl Grey of Fallodon, W. H. Hudson, Peter Scott, Lord Alanbrooke and the people who feed the pigeons in Trafalgar Square.

As the subject of West Indian birds is not without its sensational aspects, one must hope that Mr. Bond has seen fit to preserve a decent discretion, particularly in his treatment of the nuptial plumage of the copper-rumped hummingbird (*Amazilia tobaci*) and the private life of the scaly-breasted thrasher.

P.S. Terrible mistake! I now find out that the author of *Birds of the West Indies* is a different James Bond, Curator at the Academy of Natural Sciences, Philadelphia, and a top banana in ornithology. You may have complete confidence in this Mr. Bond, who knows all about species, habitat, nidification and the reversible outer toe of the osprey or fish hawk.

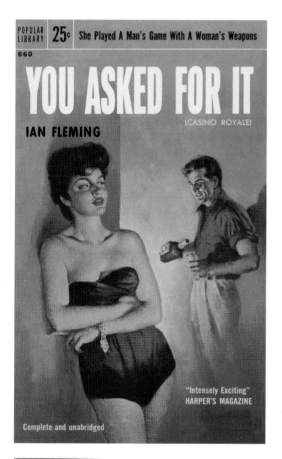

When *Casino Royale* was published as a US paperback in 1955, it was retitled. *Courtesy of J. Kingston Pierce of Killer Covers*

Julia Blakely of the Rare Book Department in the Smithsonian Institution suspected the reviewer was Fleming himself. "The review is in the same 'voice' as his letters," she wrote on a 2016 Smithsonian blog. "In his post-WWII career (Naval Intelligence), Fleming worked for the newspaper group that owned the *Times*. As its Foreign Manager, he was responsible for its stable of correspondents from 1945, working full-time until late in 1959 although he continued to be a contributor until 1961."

Fleming biographer Matthew Parker noted in *Goldeneye* that as far back as the publication of *Casino Royale*, "as a journalist, Fleming was well placed to encourage reviews of his book. His own paper, the *Sunday Times*, declared him "the best new thriller writer since Eric Ambler."

It is no stretch to think the playful Fleming was having one of his little jokes. If he was indeed the reviewer, it was an appropriately covert and subtle way for him to let the metaphorical cat out of the bag, since sooner or later Jim Bond was apt to see a copy of the review.

Bond and his wife did read the review and were baffled. As she wrote in *How 007 Got His Name*, "We could not make head nor tail of it. Why this facetious approach to the rarely hilarious profession of ornithology? And this curious reference to a collector as a bird-watcher? And in the *Sunday Times* of all places."

It so happened that his publisher's overseas representative was in Manhattan at the time, and he explained to the Bonds that an author named Ian Fleming had written several thrillers that featured a secret agent named James Bond. According to Mary Bond, that was the first time they had heard of Ian Fleming.

The details of the connection—including the fact that almost nine years elapsed before Bond found out his name had been stolen—have gone largely unexamined. In various interviews, Fleming offered slightly different reasons for his choice of the name. In an article in the February 1961 issue of *Rogue* magazine, Fleming divulged the origin of his secret agent's name for the first time.

"There really is a James Bond, you know, but he's an American ornithologist not a secret agent," he said. "I'd read a book of his and when I was casting about for a natural sounding name for my hero, I recalled the book and lifted the author's name outright."

Soon after Mary discovered that Fleming had stolen her husband's name for a sexy fictional secret agent, young women started calling the Bonds late at night and asking if James was there. As Mary wrote in *To James Bond with Love*, she would answer and ask who was calling. When the young woman replied, in a seductive voice, "I think he'll know," Mary said: "Yes, James is here, but this is Pussy Galore, and he's busy now."

After fielding several such calls, she wrote to the British author in 1961: "I tell my JB he could sue you for defamation of character but he regards the whole thing as a joke."

TEN KENNEDY FAVORITES

MELBOURNE by David Cecil	THE PRICE OF UNION by Herbert Agar
MONTROSE by John Buchan	JOHN C. CALHOUN by Margaret L. Coit
MARLBOROUGH by Winston S. Churchill	BYRON IN ITALY by Peter Quennell
JOHN QUINCY ADAMS by Samuel Flagg Bemis	FROM RUSSIA WITH LOVE by Ian Fleming
THE EMERGENCE OF LINCOLN by Allan Nevins	THE RED AND THE BLACK by M. de Stendhal

Ian Fleming's James Bond caught Americans' attention when President John F. Kennedy told *Life* magazine that *From Russia with Love* was one of his favorite books. *Author's collection*

In his reply to the Bonds, Fleming explained that he wanted to give his secret agent an anonymous-sounding name, the opposite "of the kind of 'Peregrine Carruthers' whom one meets in this type of fiction." Fleming added that "one of my bibles was and still is *Birds of the West Indies* by James Bond, and it struck me that this name, brief, unromantic and yet very masculine, was just what I needed and so James II was born."

He offered a similar explanation in a 1962 *New Yorker* interview. "One of the bibles of my youth was *Birds of the West Indies*, by James Bond, a noted ornithologist, and when I was casting about for a name for my protagonist I thought, My God, that's the dullest name I've ever heard, so I appropriated it," he says. "Now the dullest name in the world has become an exciting one."

Fleming also told the interviewer from the CBC on a February morning in 1964 that "when I started to write these books in 1952, I wanted a really quiet, flat name, and one of my bibles out here is James Bond's *Birds of the West Indies*, which is a very famous ornithological book indeed. I thought, 'James Bond, that's a pretty quiet name,' and so I simply stole it and used it."

This latter version was corroborated by his lifelong friend Ivar Bryce, who wrote in *You Only Live Once* that Fleming confided to him that "he was planning his first book—an intention that had been in his mind since boyhood—and the plot was already clear in his mind. It would be a spy story with, as the hero, a British agent who would beat the toughest pros that the Soviets, the Germans or the [Japanese] could muster. Ian had chosen the hero's name already, although the title of the book was still to come. He had wanted a straightforward English name, simple and solid, with no overtones to distract the reader. He had the perfect one, he said, just what he wanted: it could be seen in Goldeneye's big room, as visible as the blue-painted floor and as the fine set of Manège prints of gavottes and caprioles that lines the walls.

"I looked around at the familiar furniture, solid chairs and table built of green mahoe (a tall tree strangely related to the hibiscus) by a local carpenter to Ian's specifications, the shell collection and the piles of books. 'Look at the top,' he said, pointing to a small handbook entitled *Birds of the West Indies* which we often needed to consult. 'Look—the name,' *Birds of the West Indies* by James Bond."

Bryce's reply: "Yes, that would make a pretty good name."

A curious coincidence, perhaps, is that in Britain, "birdwatcher" is slang for "spy." What better name for Fleming's fictional secret agent than that of a real-life ornithologist (see chapter 008) or a real-life secret agent (see chapter 009)?

In 1961, less than two months into John F. Kennedy's presidency, *Life* magazine published a lengthy feature about the youthful, charismatic new leader's reading habits, which could be described in one word: voracious. The article featured a list of "ten Kennedy favorites" in the following order: *Melbourne* by David Cecil, *Montrose* by John Buchan, *Marlborough* by Winston S. Churchill, *John Quincy Adams* by Samuel Flagg Bemis, *The Emergence of Lincoln* by Allan Nevins, *The Price of Union* by Herbert Agar, *John C. Calhoun* by Margaret L. Coit, *Byron in Italy* by Peter Quennell, *From Russia with Love* by Ian Fleming, and *The Red and the Black* by M. de Stendhal.

Chapter 006

What's striking about the *Life* list is that nine of the books are scholarly works better suited for a syllabus than a sitting room. The fact that just one work of popular fiction, *From Russia with Love*, was on the list drew that much more attention to the spy novel. The American public's infatuation with 007 had begun, and Jim and Mary Bond braced for the worst. She started reading Fleming's thrillers in an effort to understand what was going on. The perpetually proper Jim couldn't get past the steamy *Dr. No*, but Mary devoured all of them. "I was forcibly struck by a marked similarity of many West Indian bars, waterfronts, personalities and even incidents described by Ian Fleming, to those Jim had related to me as his own experiences," she explained in *How 007 Got His Name*.

Refusing to let the matter rest, Mary immediately sent the previously mentioned letter to Fleming via his publisher in London:

Feb. 1, 1961

Dear Mr. Fleming,

It was inevitable that we should catch up to you! First, the review in the Sunday Times of my husband's new Birds of the West Indies revealed the existence of James Bond, British Agent.

Second, our friend Charles Chaplin of Haverford, Pa., and a friend of your brother Peter, gave us a copy of *Dr. No*, which explained the rest.

I read further stories about JBBA and became convinced that you must have been following JB *authenticus* around the West Indies, and picking up some of his adventures. It came to him as a surprise when we discovered in an interview in *Rogue* magazine, that you had brazenly taken the name of a real human being for your rascal! And after reading *Dr. No*, my JB thought you had been to Dirty Dick's in Nassau and talked with Old Farrington and got from him the story about the "Priscilla" and a wild trip about Jim's collecting parrots on Abaco. That was the time he spent several nights in a cave full of bats to get away from the mosquitoes.

As a rule truth is stranger than fiction but your JBBA proves this isn't necessarily so! Just don't let 007 marry—at least until he is 55!

This is a hurried letter because we're getting off to the Yucatan and Cozumel this afternoon, then back to Nassau where we'll spend a few days with the Chaplins.

I tell my JB he could sue you for defamation of character but he regards the whole thing as a joke.

Sincerely yours,
Mary Wickham Bond

She received a prompt reply from Fleming's assistant, addressed to her in Chestnut Hill, Philadelphia, Pasadena, USA (presumably, his assistant did not know what "Pa." stands for).

Dear Mrs. Bond,

As Mr. Fleming is abroad, I write to acknowledge your letter to him. I know that when he returns he will be most interested to read your letter and to hear some of the adventures of JB *authenticus* to whom, may I add, judging from the photograph [most likely referring to a photo of Mary Bond's photo of her husband skinning a bird on the back of the dust jacket of the 1960 edition of *Birds of the West Indies*], JBBA bears a striking resemblance.

Yours sincerely,

Beryl Griffie-Williams

A few months later, in a letter dated June 20, 1961, Fleming himself responded from his Fleet Street address in London. In the letter, addressed to "Mrs. James Bond," Fleming fessed up to stealing her husband's name and made three generous offers.

He gave the real James Bond "unlimited use of the name Ian Fleming for any purpose he may think fit," and to discover "a horrible new species" and "christen insulting fashion" as "a way of getting his back!"

Fleming also offered to let the Bonds stay at Goldeneye so that they could inspect in comfort "the shrine where the second James Bond was born."

For the rest of his life, the real James Bond would be increasingly overshadowed by an amoral fictional spy. As the 007 craze grew, the Bonds started getting more and more wisecracks thrown their way, to the point where Mary Bond kept track of them in her memoirs. Hotel bellboys winked. Taxi drivers cracked jokes. Folks at cocktail parties gushed. And a publicity-minded cinema owner on Philadelphia's Main Line offered Bond $100 to arrive by helicopter in front of his movie theater for the local premiere of *Goldfinger*. The answer: a flat no.

When the Bonds traveled to England just a few months after Fleming's letter, they were greeted by the customs officer with a "Well, well, well. James Bond—in person!" In London, Mary opened the latest edition of the *Daily Express* and found a James Bond comic strip.

The Philadelphia Chewing Gum Company soon marketed James Bond trading cards that

The Philadelphia Chewing Gum Co. marketed James Bond trading cards featuring 007, not the ornithologist. *Courtesy of www.toysofbond.co.uk*

featured 007, and not the hometown ornithologist of the same name. As Mary wrote in *How 007 Got His Name*, she and her husband were "more or less resigned to the element of fantasy that dogged our lives."

The one time that Jim Bond seemed to enjoy the 007 connection came in early 1965, after Bond tracked down a bird thought to have been hunted into extinction—in a freezer in Barbados. Mary wrote that a hunter had shot it on the Caribbean island, thought it odd looking, and had given it to a birdwatching friend of Bond's, who put it in his fridge and notified the ornithologist.

When the friend, Maurice Hutt, had telephoned and said, " I think I have something rather interesting to show you," Bond figured it had to be a valuable find, but couldn't imagine what. The moment Bond saw the dead bird in Barbados months later, he immediately knew it was an exceedingly rare Eskimo curlew.

The foot-long shorebird had once been abundant in the Western Hemisphere, where it nested on the Arctic tundra in the summer and migrated in flocks numbering hundreds of thousands of birds to the southern tip of South America for the winter. But the Eskimo

Newspapers had a field day when Bond uncovered an Eskimo curlew, previously thought to be extinct.
Free Library of Philadelphia, Rare Book Department

curlew was also a popular game bird for market hunters, and by the late 1800s it had been deemed "a vanishing race—on the way to extinction." Habitat loss was also a key to its extinction, and by the early 1960s, Eskimo curlews had virtually disappeared.

At the time, Bond's discovery was considered the final curfew for the Eskimo curlew. Since then, there have been sightings of a flock of twenty-three in Texas in 1981 and a lone bird in Nebraska in 1987.

The story of Bond's discovery, which surfaced around the time of the 007 movie *Thunderball*, made the news all over the United States and beyond—no doubt in large part to the name game. Headline writers had a field day: "James Bond Finds Clue to Curlew Killer," "James Bond in Case of the Vanishing Curlew," "This James Bond Catches Birds Instead of Villains," and so forth.

Sports Illustrated even ran an item in its May 24, 1965, edition:

> After a long stalking of his quarry James Bond killed quickly, silently and patiently. Then, in a variation of techniques not yet screened at any neighborhood movie theatre, Bond skinned his victim and sent him back to headquarters to be stuffed. "A hundred years from now, this may be the last known specimen of the Eskimo Curlew," crowed Bond, Curator of Birds at the Academy of Natural Sciences of Philadelphia. He hadn't been so excited since Ian Fleming borrowed his name to christen Secret Agent 007.

Soon after, Bond received an angry letter from a Yale biology student who had read the *Sports Illustrated* squib. The undergrad wrote that he was "horrified that an eminent ornithologist such as you would even think of shooting such a bird, even if your colleagues would not believe you saw an Eskimo Curlew. This type of activity harkens back to the 19th century when 'biology' was the wholesale slaughter of animals to fill museum collections."

When Bond explained to the student that *Sports Illustrated* had played fast and loose with actual events, the student sent an apology, as recounted in *How 007 Got His Name*: "In my youthful zeal I accepted as factual, information from which I should have realized as being a source of distorted facts."

Later in 1964, the movie *Thunderball* debuted. Sean Connery again played James Bond, swimming just below the water surface with a fake gull on his head as a disguise.

That same month, at the annual meeting of the American Ornithologists' Union, organizers distributed a parody of the group's newsletter, *The Auk*. The publication was the *Auklet*, billed as an "Irregular Journal of Irreverence," featured "On Her Majesty's Ornithological Service," by Avian Flemish.

The parody, set in the West Indies, pits British ornithological agent James Blond against avian archvillain Goldfincher, who has a dastardly plot to collect as many endemic birds as he can, then drive up their value by blanketing the region with radioactive birdseed. (To read the parody, see appendix 2.)

As Mary noted in *To James Bond with Love*, "The trouble was Fleming had stepped out of the picture and had left Jim holding the bag, and Jim wasn't half as interested in getting some of his own [stature] back as in being left completely out of the limelight."

Bond had little time for the 007 nonsense. But for a while at least, his wife loved the attention. Insisting that her husband was more handsome than 007 actor Sean Connery, who was more than thirty years younger, Mary became Jim's biggest publicist.

Chapter 006

The Love of His Life

James and Mary Bond married in 1953, when both were in their fifties. They first met in the 1930s in Philadelphia while Mary was researching an article for *Audubon* magazine, according to David Contosta's *Private Life of James Bond*. After Mary's first husband died in 1952, she took note of Bond, who had inherited a little money in the 1940s and purchased a convertible that he drove to Maine in the warmer months.

"He was the mystery man of Mt. Desert to all the young girls who summered in Maine," Mary Bond told *Philadelphia Bulletin* interviewer Pete Martin in 1964. "Here was this handsome creature dashing about in a red automobile, and a lot of people, mostly of the female persuasion, wanted to know who he was. He appeared and went like a Cheshire Cat."

A published novelist and poet, she became Jim Bond's unofficial publicist once the 007 connection kicked into high gear. She wrote three books about Bond and their travels together, including *How 007 Got His Name*.

The publisher of Mary Bond's first book, *How 007 Got His Name*, was Collins, the same house that published Fleming's later 007 novels. The front of the dust jacket featured a revolver with four bird feathers by popular cover illustrator Arthur Barbosa, after Richard Chopping's designs for the dust jackets of the original 007 novels. The back of the dust jacket was a copy of the page that Fleming had inscribed to Jim Bond two years earlier. The dedication played off the inscription: "To the real James Bond—a quiet man with a quiet name."

The book was published more than fifty years ago, when 007 was becoming a cult figure, and Mary relished the spotlight. She later wrote *Far Afield in the Caribbean* (1971) and *To James Bond with Love* (1980). On a trip to England to promote the latter book, she discussed the similarities between her husband and 007 actor Sean Connery. "They have some similarities, some similar interests. They like birds. But my husband is different. He shoots the birds. And of course they have a similar interest in guns and have their own special kinds. And I think the shaking of martinis or the stirring are very important similarities."

The Bond-Fleming Bond

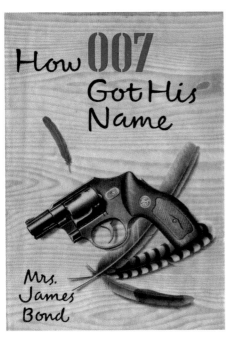

Mary Wickham Bond couldn't resist capitalizing on her husband's name, to his chagrin. *Author's collection*

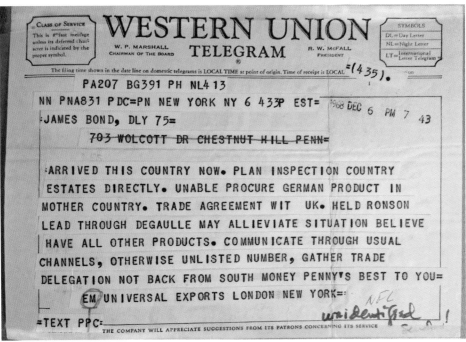

Jim and Mary Bond started receiving anonymous gag telegrams such as the one above. *Free Library of Philadelphia, Rare Book Department*

Chapter 006

The Real James Bond's Martini
"I Just Let Fly with the Gin"

In *Dr. No*, the fictional James Bond ordered a medium vodka dry martini with a slice of lemon peel, shaken and not stirred, with preferably Russian or Polish vodka. The real James Bond had his own recipe. "I don't measure it out, jigger by jigger," he told Dan Thomas of the *Sunday Bulletin Magazine*. "I just let fly with the gin and in the end I give it just a touch of vermouth."

Bond agreed that the ratio was roughly six parts of gin to one part of vermouth, adding: "I do rub a little lemon peel on the rim, and the glass should be ice cold."

Mary Bond told the interviewer that when her husband made a martini, it was a ceremony.

"I see no point in not making it right," Bond added. "A poor martini is one of the worst drinks in the world. If it's too sweet and too warm, it's awful."

The real Bond did not specify if the cocktail should be shaken or stirred.

By 1980, however, when *To James Bond with Love* was published, Mary Bond acknowledged the 007 connection had worn thin. "Although amusing at times, it is far from agreeable to be swept into a legend by a constant invasion of a fictional character of dubious reputation," she wrote in the preface to a book that capitalized on that very same name game.

Ironically, after Sean Connery embodied 007 to the movie screen, the fictional secret agent overshadowed Fleming too. As Fleming's widow, Ann, wrote in an article for the October 1966 *Ladies' Home Journal*: "As Ian became identified with James Bond, he somehow became more like the James Bond he had invented. . . . He was running away from himself, I suppose." The title of her article: "How James Bond Destroyed My Husband."

Observed John Pearson in *The Life of Ian Fleming*, "To give the tale its final Gothic twist, while Ian was becoming increasingly dispirited and sick, James Bond was flourishing, no longer his creator's faithful slave but in his new livelier, sexier and wittier embodiment in the young Sean Connery."

Bond even became the brunt of a *Philadelphia Evening Bulletin* reader-help columnist. *Free Library of Philadelphia, Rare Book Department*

The Bond-Fleming connection continues to live on, thanks in large part to the 007 movies. In an early scene in the 2002 Bond film *Die Another Day*, for example, actor Pierce Brosnan's 007 examines a *Birds of the West Indies* field guide in a Havana hotel. When he meets the requisite Bond girl, Jinx (played by a voluptuous Halle Berry), he makes a 007-style quip. "I'm just here for the birds," he explains. "Ornithologist."

Jumping on the Bondwagon, the ITV *Miss Marple* TV series featured a 2013 episode titled "A Caribbean Mystery," in which Agatha Christie's sleuth meets Ian Fleming at a talk on birds of the West Indies, given by James Bond. Before the talk, Fleming confides that he's working on a new book but can't come up with a name for the main character. And when Bond steps to the podium and introduces himself as "Bond, James Bond," a metaphorical lightbulb goes off in Fleming's head. For serious Fleming buffs, there's even a reference to a bit of arcana from *Dr. No*. The subject of James Bond's talk is guano, and playing the role of James Bond the ornithologist is the teleplay's screenwriter, Charlie Higson, who also authored a series of young adult books about 007 as a youth.

But for sheer creativity, it's tough to top the Cumberland Bird Observers Club in Australia, which has used the Fleming-Bond connection to promote birdwatching. In 2001, for example, the 560-member club produced a flyer featuring a photo of a dashing, binoculars-wielding birder in a tuxedo, accompanied by four women birders in sexy gowns and miniskirts. The goal: attracting new members by showing that birdwatching could involve more than just stodgy guys and gals in drab outfits gawking at ducks through spotting scopes and binoculars. The club even had a web page devoted to Jim Bond.

In an August 21, 2001, article in the *Sydney Morning Herald* headlined "Long Lenses, Short Skirts and a View to a Thrill," club member Andrew Patrick explained, "We were looking for something sexy. In the past, we've just handed the usual leaflets out." Club membership increased by 8 percent after the new flyers arrived. "We were really pleased with the response," said Patrick, whose wife was one of the "dolly-birds" in the flyer.

Jim Bond the ornithologist and James Bond the secret agent, forever linked.

British sculptor Anthony Smith created this bronze bust of Ian Fleming for the Fleming family to commemorate the centenary of Fleming's birth in 2008. *Creative Commons*

From *You Only Live Twice* by Ian Fleming, published by Jonathan Cape Ltd. *Reproduced by permission of the Random House Group Ltd.* ©1964

CHAPTER 007

You Only Live Twice

One of the lingering mysteries surrounding that day in February 1964 when Jim Bond met Ian Fleming at Goldeneye is this: What has become of that celebrated first edition, first printing, of *You Only Live Twice* that Fleming inscribed and gave to "the *real* James Bond"?

Fleming wrote this eleventh 007 novel at his Jamaica estate in 1963. Jonathan Cape first published it in Great Britain on March 26, 1964, seven weeks after Bond and Mary dropped in at Goldeneye and received an advance copy of the book. It was the final book in what has become known as the Blofeld Trilogy, named for 007 villain Ernst Stavro Blofeld. The book, which retailed for sixteen shillings (roughly $19), was the last 007 novel published while Fleming was alive.

Set mostly in Japan, the novel hews fairly closely to the standard 007 formula, with a dash of amnesia thrown in. Besides sex siren Kissy Suzuki and the villainous Blofeld, who is posing as Dr. Guntram Shatterhand, the book features a premature obituary of the fictional James Bond, which made it an interesting choice for Fleming to give to the real James Bond.

The 2008 *Financial Times* article "Buying Ian Fleming's Books for Investment" explained that this particular volume was extremely valuable for one simple reason: It's considered by some to be the ultimate in what's known in the book-collecting trade as an "association copy"—which in this case meant Fleming's signature. The article's author, Simon de Burton, wrote: "An avid book collector himself, Fleming knew the value of his own signature and was selective about handing out inscribed copies, meaning that survivors are particularly sought after, especially if they have associations. Undoubtedly the Holy Grail of such association copies is the first edition of the 1964 novel *You Only Live Twice*, which Fleming gave to an unexpected visitor who wandered up to his Jamaican island hideaway, Goldeneye, and pronounced himself to be the author of *A Field Guide to Birds of the West Indies*. His name was Bond—James Bond."

Chapter 007

Jim and Mary Bond, who were married more than thirty-five years, donated their archives to separate institutions. *Free Library of Philadelphia, Rare Book Department*

The dust jacket for the 256-page novel featured a Richard Chopping illustration of a pink chrysanthemum and a Japanese-style toad crushing a dragonfly. The background is a light-tan bamboo. The odd motif didn't seem to hurt sales. The book had 62,000 preorders and went into a second printing a month after its publication date. The title was derived from Bond's attempt at a haiku in chapter 11, where he tries to write in the style of seventeenth-century Japanese poet Matsuo Basho.

On the jacket flap, and in gilt lettering on the front of the book, are a series of symbols that supposedly represent the book's title in Japanese. According to Japanese journalist Hiroki Fukuda, "The Japanese characters read 'Nido Dake No Seimei.' It literally means 'Live Only Twice' but it doesn't make sense in Japanese. It is not an appropriate translation of 'You Only Live Twice' and moreover, the last character is not correct. 'You Only Live Twice' is actually quite hard to translate into Japanese. The Japanese title of the novel and the movie was *007 Wa Nido Shinu*, which literally means '007 Dies Twice' in English."

The book Fleming inscribed to the real James Bond traveled with the Bonds from Jamaica back to their home on Davidson Road in the Chestnut Hill section of Philadelphia, and later to their apartment on the top floor of Hill House near the Chestnut Hill commuter train station. According to Robert McCracken Peck of the Academy of Natural Sciences, Mary loved to show the book to guests. After Bond died, Mary and Peck talked many times about whether she would eventually donate it to the academy.

"She said no, no, this really has to do with the Ian Fleming side of his life," Peck recalls. "I'm going to divide up my papers in two ways. The academy will get all of his scientific papers and the Free Library of Philadelphia will get all the rest."

That had been Bond's wish as well. Thus, the inscribed copy of *You Only Live Twice*, along with several boxes filled with 007 and Bond press clippings, book reviews, and memorabilia, went to the Free Library of Philadelphia's Rare Book Department but vanished soon after it arrived.

```
From Mary Wickham Bond
To the Free Library of Philadelphia
Rare Book Department        BONDIANA
                         INDEX OF ITEMS
```
*007 Bondiana
Received at Library
April 1975*

1. Framed original letter from Ian Fleming to Mrs. James Bond
   Feb.1, 1961
2. Inscribed copy of "You Only Live Twice" presented by Ian Fleming
   to James Bond at Golden Eye, Jamaica, Feb. 5, 1964
3. Photographs, film and slides of Fleming, Bond, etc, at Golden Eye
   Feb. 5, 1964

Mary Bond initially donated the inscribed first edition of *You Only Live Twice* to the Free Library, only to secretly change her mind. *Free Library of Philadelphia, Rare Book Department*

The culprit? Mary herself. According to biographer David Contosta, "Mary had read somewhere that inscribed books like hers could be sold for a tidy sum at auction, so she got it into her head to sell the book even though she did not need the money. I couldn't convince her otherwise. Mary knew a woman who worked down there at the library. Mary said she just wanted to see the book again, and asked her to bring the book up to her [in Chestnut Hill]."

She never gave it back. According to Caitlin Goodman, curator of the library's Rare Book Department, instead of returning the book, Mary consigned it to Sotheby's auction of English Literature and History at Aeolian Hall in London on July 11, 1996.

Lot 283 took up two and a half pages in the catalog and included a promotional card with a photograph of Fleming and Bond taken when they met at Goldeneye in 1964, an enlargement of an original photograph of Bond skinning a European cuckoo in Barbados in 1989, and photocopies of Mary's book *To James Bond with Love*.

The book brought 12,500 pounds ($17,092)—considerably more than its pre-auction estimate of 6,000 to 8,000 pounds. The same auction featured a lot of eleven Fleming first editions and fifty-nine first and later editions by or about Fleming and the fictional James Bond. That lot sold for 1,150 pounds—again, much more than the pre-auction estimate of 500 to 700 pounds.

After the Sotheby's auction, the book disappeared from the public radar until 2008, when it resurfaced as part of Lot 103 at an auction of movie memorabilia. Hollywood Auction 33, held on December 11 at the Profiles in History auction house's offices in Calabasas Hills, California, featured a dozen other Fleming lots, including signed first editions of *From Russia with Love* and *Casino Royale*, plus Roald Dahl's handwritten 100-page screenplay for the film version of *You Only Live Twice*.

Other highly publicized items were a flying saucer miniature from *Forbidden Planet* (Lot 324), featured on the catalog's cover, and a prop lightsaber that Mark Hamill was said to have used as Luke Skywalker in *Star Wars: A New Hope* and *The Empire Strikes Back* (Lot 347).

Lot 103 featured the Fleming book inscribed to James Bond, with the book smartly packaged in a dust jacket in fine condition (it lacked a dust jacket in the previous auction) and a custom half-Morocco clamshell slipcase, plus two items that confirmed the book's

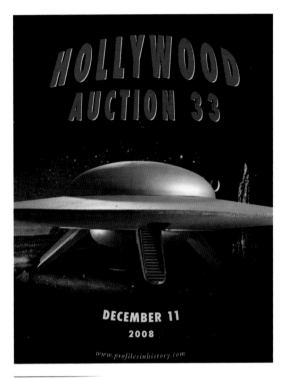

The inscribed copy last surfaced in 2008, when it was put up for sale in Profiles in History's Hollywood Auction 33. *Author's collection*

provenance and reinforced its preeminent status as an association copy: Mary Bond's *How 007 Got His Name* and a dust jacket from the original 1936 edition of James Bond's *Birds of the West Indies*. In fact, the lot took up all of page 36 in the oversized catalog and featured photos of Fleming's endpaper inscription and the front dust jackets of all three books.

The blurb for Lot 103 touted the book as "the only book ever inscribed for the real-life Bond by Fleming, who died several months after the meeting" and "the ultimate James Bond association! Fleming acknowledges the source of secret agent 007's name!"

The pre-auction estimate for the lot was $60,000 to $80,000. It sold for $72,000, plus a buyer's premium that brought the price to $82,600. How much value did that Fleming inscription add to the book? The same first British edition, first printing, from March 1964 was selling for $425 in 2017.

At that same Hollywood auction in 2008, the signed first edition of *From Russia with Love* (Lot 95) sold for $32,450 including premium, and the signed first edition of *Casino Royale* (Lot 94) fetched $20,060 including premium. Four other Ian Fleming–signed first editions, however, failed to meet their opening bids. London rare-book dealer James Pickard, who specializes in Ian Fleming first editions, says that other 007 novels are more valuable—notably the inscribed first editions of *Casino Royale*. Meanwhile, the Luke Skywalker lightsaber sold for $236,000 including premium, the highest price at the auction.

The first edition of *You Only Live Twice* that Fleming inscribed to Jim Bond has not surfaced since then. It likely sits in a safe deposit box or a book collector's dust-free, UV-light-protected bookcase—for his or her eyes only.

Also unaccounted for is Fleming's copy of the original 1936 first edition of *Birds of the West Indies* that prompted him to borrow Bond's name. Perhaps if Fleming had given it more thought, he would have included it in his extensive rare-book collection, which now resides in the University of Indiana's Lilly Library. It probably contained his personalized bookplate, which features the Fleming crest (a goat), his full name (Ian Lancaster Fleming), and the Fleming motto "Let the deed shaw."

According to tradition, these were the words that Sir Robert Fleming uttered to Robert Bruce (later King Robert I) after Fleming beheaded Bruce's chief rival at a church in Dumfries in 1306 and presented the head to Bruce. "Let the deed shaw" means "Let the deed speak for itself" or "Let the deed be manifest."

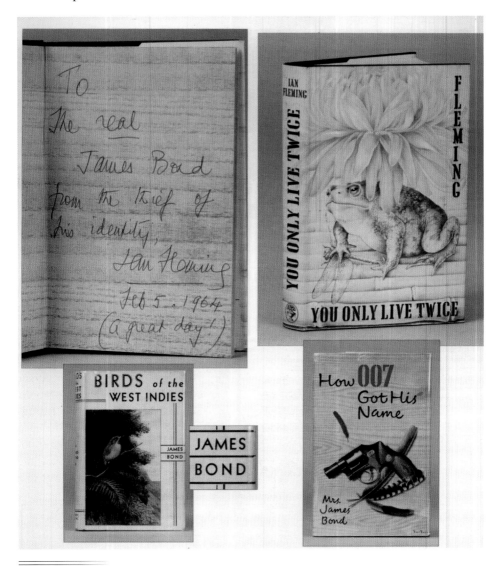

The page of the auction catalog featuring the inscribed *You Only Live Twice*. Author's collection

Richard Meinertzhagen, shown here in 1915, was a spy, ornithologist, and pathological liar. *Public domain*

Chapter 008

Twitchers & Spooks

If Ian Fleming's choice of name for his secret agent was mere happenstance, it was also serendipitous. Ornithologists and birdwatchers have been some of the most successful members of the intelligence community in modern British and American history, and some of the most notorious too.

Although ornithology as a profession took root in the early 1800s, it wasn't until the twentieth century that museums and universities routinely sent field scientists to far-flung corners of the globe. These researchers, especially ornithologists, became natural recruits for intelligence work during the two world wars, the Cold War, and other conflicts.

In a 2015 article for *The Guardian*, Helen Macdonald, author of *H Is for Hawk*, explained why birds and espionage dovetailed so nicely: "Birdwatcher is old intelligence slang for spy. . . . You have the same skills, the ability to identify, recognize, be unobtrusive, invisible, hide. You pay careful attention to your surroundings. You never feel part of the crowd."

Ornithologists who work in foreign countries have an additional skill set that includes self-reliance, some fluency in other languages, and a familiarity with the local terrain, government, and customs. They're outsiders but not strangers, so they don't draw attention. They carry surveillance equipment such as spotting scopes and high-powered binoculars for their job. And they know their way around firearms.

Hundreds of birdwatchers/spies have likely existed over the past 120 years, including such notable Brits as Harry St. John Bridger "Sinjin" Philby (1885–1960), the now-notorious Richard Meinertzhagen, and British spymasters Maxwell Knight, Martin Furnival Jones, and Andrew Parker.

Sinjin Philby was a naturalist and brilliant double agent who helped bring the Middle East's house of Ibn Saud under British influence during and after WWI, only to later help transfer the concession for Saudi Arabia's vast oil holdings to the United States—a move that some believed bordered on treason.

Richard Meinertzhagen developed a fascination for birds as a youth—even poaching them as a student at Harrow in the early 1890s. He graduated in 1895, eighteen years before Jim Bond arrived on the campus. He gained fame as the inventor of the Haversack Ruse during WWI. A tireless self-promoter, he claimed he risked his life to plant a knapsack full of British military secrets along the front lines in Palestine to trick the enemy. The false documents led the Germans and their Palestinian allies to fortify their defenses at Gaza instead of the crucial freshwater wells 26 miles inland at Beersheba.

Ian Fleming was an acquaintance of Meinertzhagen's in the 1930s and is said to have patterned some of 007 after the colorful intelligence officer. During WWII, Fleming was a commander in British naval intelligence and helped appropriate Meinertzhagen's Haversack Ruse. The plan, known as Operation Mincemeat, involved planting disinformation on a corpse dumped by submarine along the Spanish coast to make the Nazis think the Allies were invading Greece instead of Sicily.

Decades after Meinertzhagen's death, author Brian Garfield meticulously deconstructed the spy's military exploits and ornithological career in *The Meinertzhagen Mystery: The Life and Legend of a Colossal Fraud*. Meinertzhagen, it turned out, not only exaggerated and misrepresented the impact of the Haversack Ruse and his role in it, but he was also was a real-life Baron Munchausen who used his diaries and memoirs to concoct wild stories of espionage out of half-truths. He is still suspected of committing a nasty wartime atrocity or two, as well as murdering his second wife, also an ornithologist, in 1928.

Martin Furnival Jones ran MI5, the United Kingdom's internal security service, from 1965 until 1972. Andrew Parker took that position in 2013. Maxwell Knight, who specialized in supervising teams of undercover agents for several decades, is said to have been an inspiration for the character M in the 007 novels.

Knight went on to host a radio nature show for children, calling himself "Uncle Max." Off the set, he reportedly kept a blue-fronted Amazon parrot in the kitchen and a Himalayan monkey in the garden. Knight also had the perfect pet bird for a secret agent—a common cuckoo named Goo. Cuckoos, like spies, are known to pretend to be what they are not and to lay their eggs in the nests of unsuspecting hosts—a bit of avian treachery.

Knight capitalized on his celebrity status to write more than a dozen books about nature. One such opus, *Animals and Ourselves*, featured a cover photo of Knight holding a young fox. The text that followed was leavened by the lively pen and ink drawings of David Cornwell, a young nature artist and intelligence officer. You may know him better by his nom de plume, John Le Carré.

"I actually worked alongside Maxwell Knight in MI5 for a year or two, so the literary partnership is not surprising," Cornwell wrote in an email. "Our connection was of course known to our employers, but to no one else."

A more recent American example of spymaster/birdwatcher is James Schlesinger, CIA director after Watergate and later secretary of defense under Presidents Nixon and Ford. Schlesinger was known for his hard-nosed managerial style and abrasive personality. In a 2011 article for the *New York Times*, the author Richard Conniff wrote that CIA historian Nicholas Dujmovic told him, "the only nice thing I've ever heard about Schlesinger is that he was a birdwatcher." Neither Schlesinger, Jones, Parker, nor Knight, however, was known to use ornithology as a cover for spying activities. They were too high up in the pecking order to find it useful. Birdwatching was more of a diversion that helped take their mind off the grind of running a high-pressure intelligence outfit.

Then there was Fleming himself, an avid birder who learned the espionage trade as deputy to Admiral John Godfrey, director of British Naval Intelligence during WWII.

Fleming's overseas travels included trips to the United States (to coordinate the two nations' intelligence operations), to Canada (where he reportedly attended Camp X, a secret training ground for OSS agents), and to Jamaica (to attend a naval conference and, according to Matthew Parker's *Goldeneye*, to investigate rumors of a secret Nazi submarine base in the Bahamas).

In a 2008 *Times* of London article, author Ben Macintyre described Fleming's wartime communications with US general William "Wild Bill" Donovan, who was then forming the Office of Strategic Services (OSS), the forerunner of the Central Intelligence Agency. One seventy-two-page memo from Fleming in 1941 advised that the ideal secret agent "must have trained powers of observation, analysis and evaluation; absolute discretion, sobriety, devotion to duty; language and wide experience, and be aged about 40 to 50."

The OSS was created as a fourth arm of the US military, providing intelligence, propaganda, and commando operations. Donovan recruited Americans who traveled overseas, studied world affairs, and spoke foreign languages. That typically meant going after the so-called "best and the brightest" at universities, businesses, and law firms on the East Coast. The list of notable people involved with the OSS includes future Middle East peace negotiator and Nobel Peace Prize winner Ralph Bunche, future cookbook author Julia Child, future CIA director William Casey, actor Sterling Hayden, future CIA director Allen Dulles, and future Supreme Court justice Arthur Goldberg.

OSS operatives also included three of Bond's contemporaries at the Academy of Natural Sciences in Philadelphia and at least four of his fellow ornithologists at major natural-history museums. All of them knew Bond and had made ornithological expeditions to foreign countries. They also corresponded with Bond, belonged to the same professional groups, and in some cases worked alongside one another.

They included Brooke Dolan II and Frederick E. Crockett of the Academy of Natural Sciences, Herbert Girton "Bert" Deignan of the Smithsonian Institution, S. Dillon Ripley II of Yale's Peabody Museum (he had also worked at the ANSP), James Paul "Jim" Chapin of New York's American Museum of Natural History, and W. Rudyerd "Rud" Boulton of the Field Natural History Museum in Chicago. Yet another ornithologist, Emmet Reid Blake of the Field Museum, worked for US Army counterintelligence and had the most incredible espionage career of them all.

(As Gordon Corera wrote in *Operation Columba*, the OSS also employed carrier pigeons, "dropping them behind enemy lines in Europe and with agents in Asia." But that's a spy of a different feather.)

Although much of what the OSS did is still under wraps, you can get a glimpse of its activities—and the paperwork its employees had to contend with—by visiting the National Archives at College Park, Maryland. Located in an enormous glass-and-steel building near the sprawling University of Maryland campus, the 1.8-million-square-foot building is home to 21,000 boxes—7,000 cubic feet—of OSS records, including personnel files. That material ranges from expense-account paperwork and human-resources-department minutiae to an occasional nugget. The greater the achievement, the thicker the files and the more glowing the prose.

America's ornithologist spies were largely anonymous, performing highly stressful jobs in postings in such places as Africa and Asia. Spycraft was often a bad fit. According to a declassified history of the agency, *War Report: Office of Strategic Services; Operations in the Field*, communications challenges severely hampered agent operations. "SI[Secret Intelligence]/Africa first pursued the policy of selecting agents from persons who had already resided in the target areas," the report reads. "The policy was eventually abandoned when experience showed that such individuals were scarce and often not those most suited for intelligence work."

James Paul Chapin

James P. Chapin, shown here in Leopoldville, Congo, in 1942, worked for the OSS there and was a Bond contemporary. *Courtesy of TL2 project, Lukuru Foundation*

The anonymous authors of that report may well have had Jim Chapin in mind when they wrote their appraisal. Future spy Chapin (1889–1964), Bond's counterpart at the American Museum of Natural History, wrote *Birds of the Belgian Congo* in four parts over more than two decades, after gaining widespread fame for his discovery of a new species of African peacock in, of all places, the storeroom of a museum in Belgium.

In 1919, while studying at Columbia University, the twenty-year-old Chapin joined what turned out to be a six-year expedition to collect fauna and flora in the Congo basin. About two years into the trip, he saw an unusual feather in the headgear of an African tribal leader. Chapin couldn't identify the feather, except to determine it was a secondary flight feather, but he saved it for later reference. More than two decades later, while examining study skins at the Congo Museum outside Brussels, Chapin saw two discarded specimens of an unidentified peacock-like bird that were reportedly collected in the Congo. Chapin realized that the flight feather he'd collected twenty-three years earlier was from the same species. The following year Chapin led an expedition to the Congo to find the mystery bird. He did and called it the Congo peacock, *Afropavo congensi*.

Bond and Chapin were well acquainted. In her memoir *To James Bond with Love*, Mary writes of how Bond rushed to the American Museum of Natural History after an expedition in which he had collected several rare Saint Lucia finches, which resemble one of Darwin's famous Galápagos finches:

"On his return to New York he again stopped in at the American Museum. James Chapin, head of the African Department, was standing on a ladder working on his collection and asked Jim politely, 'And how was your trip?' Jim couldn't resist sounding a little triumphant when he replied, 'I collected a genus you haven't got in the American Museum,' and told his story. Chapin rushed off with the news to Frank Chapman, head of the bird department, who sat down then and there and wrote to Witmer Stone, curator of birds at the Academy of Natural Sciences of Philadelphia, asking for one of the specimens."

In 1942, Chapin joined the OSS and was deployed as a special assistant to the American Consul in Leopoldville (now known as Kinshasa), the capital of the Belgian Congo (Democratic Republic of Congo). He was even assigned a code name, CRISP. Chapin was given the job for two reasons—because he had spent all those years in the Congo as an ornithologist, accumulating invaluable local knowledge, and because the Congo held by far the world's largest supply of high-grade uranium.

Three years earlier, Albert Einstein had written to President Franklin D. Roosevelt warning that Nazi Germany might develop an atomic bomb and urging him to protect the uranium from the Congo's Shinkolobwe mine. The OSS wanted to get the uranium out of the mines at Shinkolobwe before the Nazis could get their hands on it. As it turned out, the mine's uranium was crucial to the Americans' Manhattan Project.

Alas, Chapin's stint undercover in the Congo did not go well. His assignment included sharing intelligence with OSS operations in neighboring countries, investigating activities by enemy agents, and expanding intelligence operations in Africa. He was soon overwhelmed by the job, which also included the painstaking chore of translating all his correspondence into secret code, which took hours at a time.

Much of Chapin's thin OSS personnel file involves medical matters, although one expense report for his Belgian Congo work included "two suits of 'safari' clothes," as well as a gun permit for a 12-gauge shotgun. By April 1943, it had become clear in Washington that Chapin couldn't handle the pressure, and he was sent stateside, where he spent several months at Johns Hopkins Hospital in Baltimore before being discharged.

Jim Chapin's ornithological expertise came in handy on his way to Africa for the OSS—forcing sooty terns to nest away from an airstrip. *Author's collection*

In September 1943, a week after being discharged, he returned to the American Museum of Natural History. After the war, Chapin and his wife, Ruth Trimble Chapin, lived in the eastern Congo from 1953 to 1958, where they studied birds. He died at his home on the Upper West Side of Manhattan on April 7, 1964, age seventy-four. At that time, he was the museum's curator emeritus, still engaged in bird research.

Waging War on the Wideawakes

On Jim Chapin's way to the Congo, his birding acumen unexpectedly proved pivotal. As Susan Williams recounted in *Spies in the Congo: The Race for the Ore That Built the Atomic Bomb*, his plane had to make a refueling stop at a US Army Air Forces landing strip on Ascension Island in the South Atlantic. When the base commander learned that an ornithologist was visiting, he asked Chapin to solve a problem that was threatening all outbound flights: a colony of sooty terns.

The birds, known locally as wideawakes, had established a major nesting ground just beyond the end of the newly constructed runway. When a plane headed down the runway, the rumble of the engines spooked the birds and sent them into the flight path. As *The Army Air Forces in World War II* (vol. 7) described the situation, heavier planes couldn't gain altitude fast enough to avoid the birds and had to fly right through them, "running the risk of a broken windshield, a dented leading edge, or a bird wedged in engine or air scoop." The problem was so bad that the airstrip was known as Wideawake Airfield.

Ground crews had tried everything from smoke candles and dynamite to feral cats, with no success. Enter Chapin, who advised that if the wideawakes' eggs were destroyed, the birds would nest elsewhere. Some 40,000 smashed eggs later, the terns relocated, and pilots and crews rested easier for a little while.

W. Rudyerd Boulton

Rud Boulton (1901–1983), a museum ornithologist in New York, Pittsburgh, and Chicago, was a specialist in the birds of Angola. He participated in several expeditions to Africa and the Americas but published little, aside from *Traveling with the Birds*, a delightful and plainspoken 1933 children's book about bird migration.

With the onset of WWII, Boulton left his job at the Field Museum of Natural History in Chicago and joined the OSS. He became head of the Secret Intelligence desk for Africa, including Jim Chapin's sector, and was connected to the top-secret import of Congolese uranium for the Manhattan Project atomic bomb development. Another prime mission was to keep tabs on German operatives in West Africa and prevent the Nazis from smuggling that uranium out of the Congo.

Boulton's OSS files include a memo instructing him (code name: NYANZA) to travel to Africa to look into counterintelligence operations. "You will complete the mission with all possible dispatch and return to Washington where you will make a complete and

Twitchers & Spooks

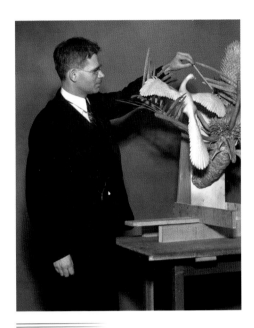

W. Rudyerd Boulton of Chicago's Field Museum of Natural History worked for the OSS as head of the Secret Intelligence desk for Africa. © *The Field Museum*

detailed report to the acting chief of X-2 and the acting chief of SI." ("X-2" was the OSS abbreviation for counterintelligence, and "SI" was special intelligence.)

Among his assignments: go to Tangier to discuss the German-Spanish intelligence situation. The OSS had learned that the Germans weren't liked and that the intelligence they gathered from Spain was the result of bribes. The OSS's plan was simple: "It is suggested we get a Spanish American of attractive personality and outbid the Germans." (The OSS was known for outbribing other intelligence agencies, friend or foe.)

After the war, Boulton apparently was not as eager to get back to museum work as Chapin. In *A Look over My Shoulder*, Richard Helms wrote about the transition from the OSS to the CIA: "The morning the termination order was announced at General Donovan's staff meeting, Rudyerd Boulton, a soft-spoken internationally known ornithologist specializing in Africa, shot up from his chair. Thrusting his arms toward heaven, he shouted: 'Jesus H. Christ. I suppose this means that it's back to those goddamn birds,' and stumbled from the room. In those days, African specialists were hard to come by, and the professor was to remain with the CIA until his retirement."

As Susan Williams recounted in *Spies in the Congo*, when Boulton learned that his former OSS Africa operative Jim Chapin was scheduled to talk to the Explorers Club in Manhattan about his experiences on Ascension Island, he wrote to Chapin about keeping his wartime stint in the Congo a secret: "No mention of the OSS or of me [should] be made during the course of it."

Two years later, Chapin named a subspecies of Cameroon scrub warbler after his OSS boss—*Bradypterus lopezi boultoni*. After retiring in the 1960s, Boulton returned to Africa with plans to conduct bird research, although the parabolic microphones he brought with him seemed better suited for eavesdropping.

The OSS efforts in Africa were successful. As Williams noted in *Spies in the Congo*: "None of the Congolese ore, so far as is known, was sent from the Congo to Nazi Germany between 1943 and 1945, whether by smuggling or any other means. And without sufficient uranium ore of that uniquely rich quality, the German atomic project could not succeed."

S. Dillon Ripley II

S. Dillon Ripley II (second from right) worked for the OSS in Southeast Asia. His duties included training Indonesian spies. *Courtesy of Smithsonian Institution Archives, image #SIA2010-0866*

Dillon Ripley (1913–2001) attended St. Paul's School in New Hampshire thirteen years after Bond. Among his achievements there, Ripley founded the Offal-Eating Club, whose members retrieved, cooked, and dined on roadkill rabbits and pheasants. Ripley later studied ornithology at Harvard, Yale, and Columbia. In 1937, the 6-foot-3 Ripley went on an eighteen-month expedition for the Academy of Natural Sciences, recruited by Bond's longtime colleague Rodolphe De Schauensee. Ripley traveled on a 60-foot schooner from Philadelphia through the Panama Canal and the South Pacific to Dutch New Guinea to collect specimens of exotic birds. (Also on the trip was ANSP contemporary Frederick E. Crockett, who also joined the OSS—but more on him later.) Ripley wrote about the expedition in his 1942 book *Trail of the Money Bird*, billed on the dust jacket as "30,000 miles of adventure with a naturalist." In his foreword, Ripley thanked several people he met in New Guinea, noting that "since I have known them much has happened. Many of them are in peril for their lives."

In 1984, an interviewer for the *New York Times* asked Ripley if he used the disguise of birdwatcher to spy on the Japanese. His reply: "Curiously enough, the British, and I suppose the Indians, Pakistanis, Ceylonese and so on, thought that it was such a marvelous part of an old-fashioned cover. Their theory was that most obviously we were spies. It never seemed to be realistic because I never could discover what someone out in the bushes could discover in the way of secrets."

Those were the words of a spy dissembling. Ripley was a fearless spy, stationed in India and Pakistan. A superior's handwritten note in Ripley's OSS file reads: "Dr. Ripley's professional planning, knowledge, consistency in foresight and sound judgment in Intelligence contributed in a great measure to the successful planning and completion of clandestine military operations against the enemy. His record of achievement, willingness to serve at advance bases and behind enemy lines in close contact with the enemy without any regard for his own safety is an inspiration to others."

Here's how a 1950 profile of Ripley in the *New Yorker* described his years as spymaster/ornithologist, all in one lollapalooza of a sentence: "Ripley, who was attached, as a civilian, to a military unit of the South East Asia Command, and whose duties included training, equipping, briefing, and sending into Java and Sumatra four Indonesian spies, all of whom were eventually killed, as well as dispatching behind-the-lines undercover men to assist guerrillas in South Burma, Malaysia, and Thailand, went on to report the sighting of such creatures as purple sunbirds, seven-sisters babblers, white-breasted kingfishers, scops owls, paradise fly-catchers, and a variety of storks and herons."

During a wartime OSS trip to the island nation of Ceylon (now Sri Lanka), Ripley met Mary Livingston, his future wife, who also worked for the OSS—as did her roommate, Julia Child.

After the war, Ripley returned to ornithology and became director of the Peabody Museum at Yale. He later served as head of the Smithsonian until 1984, a key figure in the institution's golden age of expansion and acquisitions. Ripley remained friends with Bond and went on an expedition to the West Indies with him in 1960.

At least one bird species is named for Ripley—the Himalayan shrike-babbler (*Pteruthius ripleyi*)—as well as several subspecies, including a bay owl (*Phodilus badius ripleyi*) and a Thailand pitta (*Pitta guajana ripleyi*).

Herbert Girton Deignan

Herb Deignan (1906–1968) was an OSS colleague of Ripley's in Ceylon and a 1928 Princeton grad who taught English in Thailand for four years and collected birds there in his spare time before joining the Smithsonian Institution. In 1937, several years before Deignan joined the OSS, Bond's colleague, Rodolphe De Schauensee, named a bird for Deignan in honor of his ornithological work in Thailand.

When WWII arrived, Deignan's familiarity with Asia and his ability to speak Thai made him a perfect OSS recruit. By then Japan occupied Thailand, and Deignan was assigned to Ceylon, where he worked with a group of Thais and Americans involved in the Free Thai movement. For his wartime efforts, Deignan received a medal of merit and the Order of the White Elephant (fifth class) by the King of Thailand (the first four classes of the order were reserved for royalty).

Deignan was the author of *The Birds of Northern Thailand*, published in 1945, the final year of the war. He sent back 143 bird skins, 6 bird skeletons, and 236 shells and

Herbert Deignan worked for the OSS with Dillon Ripley in Southeast Asia. *Courtesy of Smithsonian Institution Archives, image #SIA2009-1775*

barnacles, plus mammal, reptile, insect, snail, and worm specimens, to the Smithsonian that year. He was an expert in several families of Asian birds, including the Asiatic bulbuls and the babbling thrushes.

When asked on one OSS form called "Individual's personal desire as to future assignment," Deignan wrote: "To return to the Smithsonian Institution in January," which he did.

In 1952, he returned to Thailand as a Guggenheim fellow. Like his OSS colleague Jim Chapin, his happiest times in the years after the war were working as an ornithologist in museums. As his obituary in the October 1989 *The Auk* stated: "Museums and research were to him havens in a troubled world, and as he saw little in the way of easing the troubles, he preferred to absent himself from as much of them as he could."

Deignan's name is best remembered in Sri Lanka for the threatened species *Lankascincus deignani*, a.k.a. Deignan's tree skink, found in the Central Highlands of Sri Lanka 600–2,300 meters (1,968–7,544 ft.) above sea level. In 1951, Deignan named a subspecies of cocoa thrush after Jim Bond, *Turdus fumigatus bondi*.

Brooke Dolan II

Bond's ANSP contemporary Brooke Dolan (1908–1945) was an explorer and OSS member who became famous for his wartime exploits. Educated at St. Paul's School (he arrived in 1921, eight years after Bond left), Princeton, and Harvard, the independently wealthy Dolan spoke Tibetan and Chinese and was well versed in Buddhism.

Dolan first gained fame in 1931, when he led an expedition to northeastern Tibet and western China that brought the first specimens of the giant panda to the United States. During a second expedition, in 1934–1936, he collected 3,000 birds and 140 mammals from Tibet and western China and covered approximately 200,000 square miles of territories often ruled by competing warlords. The expedition members encountered infectious diseases, blizzards, quicksand, food shortages, and marauding nomads. These credentials made him perfect for the OSS.

Brooke Dolan II visited Tibet three times, including an expedition for the OSS in 1942–43. *ANSP Archives Collection 64 (restricted), Dolan, Brooke, II, 1909–1945, Papers, 1931–1946*

In 1942, President Franklin D. Roosevelt sent Dolan and Captain Ilya Tolstoy (grandson of the novelist) on a reconnaissance mission to Tibet and China to look into the feasibility of a new supply route over the Himalayas into Tibet and China to investigate possible airfield sites, and to gauge attitudes toward the war within the various regions they explored.

Here's how one of Dolan's OSS superiors, Colonel M. Preston Goodfellow, described the seven-month trek: "Throughout the accomplishment of the mission, Captain Dolan met and overcame extreme difficulties of travel, weather, terrain, excessively high altitudes and danger from native populations whose attitude toward himself and his mission was not previously known. . . . [He was able to] proceed to and through places never seen before by white men."

Along the way, Dolan and Tolstoy met with the Dalai Lama, only seven years old at the time, to give him gifts and a message from President Roosevelt.

Given the conditions, Dolan had by far the best requisition orders / expense-account items in his OSS personnel files. One equipment request included hand grenades, irritant gas, helmets, and two .45-caliber automatic Colt pistols with magazine clips and holsters.

After returning home, Dolan was transferred to the US Army Air Forces. He returned to China once more to gather intelligence for the headquarters of the 20th Bomber Command. He died there in August 1945 at age thirty-seven. The war had already ended.

Dolan's eared pheasant (*Crossoptilon crossoptilon dolani*) was named for him.

Frederick Eugene Crockett

Another celebrated contemporary at the Academy of Natural Sciences who joined the OSS was Harvard-educated Freddie Crockett (1907–1978), a dog handler and radio operator on Admiral Richard Byrd's first South Pole expedition (salary: $1 per annum).

From 1934 to 1936, Crockett prospected for gold in the southern United States and Mexico before joining the academy. Crockett's claim to fame there: he and his wife, anthropologist Charis Denison Crockett, led an expedition to the Galápagos, Polynesia, the Solomons, New Britain, and New Guinea that included fellow future OSS member Dillon Ripley.

During WWII, Crockett served in the US Navy and Army Air Corps in Greenland and the Arctic, focusing on intelligence work and sea rescues before becoming an Arctic-related military instructor stateside.

In 1945, Crockett was recruited by the OSS and sent to Jakarta on a mission codenamed ICEBERG—perhaps a nod to his days in the Arctic and Antarctic. According to William J. Rust in the March 2016 article "Operation ICEBERG" for *Studies in Intelligence*, Crockett's mission had two goals: "The first was immediate and overt: helping rescue US POWs from Japanese camps. This humanitarian assignment provided cover for a second, longer-term objective: establishing a field station for espionage in what would become the nation of Indonesia."

Chapter 008

Explorer Frederick Crockett sailed to New Guinea aboard *Chiva* in 1937 with Dillon Ripley. As part the OSS, he later rescued American prisoners of war in Indonesia. *Author's collection*

In terms of his qualifications, the assigning officer wrote that Crockett had spent a lot of time in the area, "speaks the language, and knows and understands the people. He has extensive small-boat training as well as cryptography, radio communications and procedure."

Alas, the operation's results were mixed, with Crockett reporting that his efforts to evacuate American prisoners of war and civilian internees was "directly contrary to the policy of the British and Dutch." Similarly, Crockett reported that his intelligence-gathering efforts were constantly undermined by his Dutch and British counterparts, who "were worried that US observers would report unfavorably, even though accurately, on their subtle endeavours to restore a 'status quo ante bellum.'"

According to that official history of the CIA, "[Crockett's] postwar career included an unsuccessful bid for political office in California and a return to the CIA in the early 1950s. He died in 1978, having spent the last twenty-four years of his life as a commercial real estate broker."

Admiral Byrd named Mt. Crockett, a prominent peak in Antarctica's Queen Maud Mountains, for him. In 1939, Rodolphe De Schauensee and Ernst Mayr also named a yellow-gaped honeyeater after him—*Meliphaga flavirictus crockettorum*.

Emmet Reid Blake

For pure 007-style derring-do, it's tough to top Bob Blake (1908–1997), Bond's other counterpart at Chicago's Field Museum of Natural History. Blake grew up precocious in South Carolina, with a fascination for birds and a penchant (like Maxwell Knight) for carrying reptiles in his pockets, which earned him the nickname Snakey.

He entered the Presbyterian College of South Carolina as a fifteen-year-old and soon turned a vacant dormitory into a natural-history museum for small mammals and birds that Blake and a pal had shot and skinned. Although he liked studying ornithology, he wrote to his mother that he wanted to be an explorer: "I love travel, excitement, outdoor life and the beauties of nature."

After graduating, Blake roller-skated 900 miles to become a part-time grad student at the University of Pittsburgh. He took a break from his studies to join a yearlong National Geographic Society trek up the Amazon River in Brazil, where he served as a bird skinner, cook, and dishwasher.

When the rest of the expedition went home, Blake remained. Working eighteen-hour days, he collected 803 birds, 86 reptiles, and 37 mammals. One of the lizards was new to science and eventually named *Anadia blakei*.

When he returned stateside, the Great Depression was in full swing, and he paid his way through graduate school by working jobs that included private detective, gas-station attendant, and carnival boxer.

After getting his master's in zoology, Blake joined the Field Museum in 1935 and became assistant curator of birds. Blake's interest in birds got him into US Army counterintelligence in WWII. The army had originally assigned him to be a stretcher bearer, but when an intelligence officer learned that young Blake had made several field trips to South America as an ornithologist and spoke Spanish and Portuguese, he suggested sending Blake to keep an eye on Nazi sympathizers in South America.

Instead, Blake was sent to North Africa, where he worked his way north into Europe, ferreting out German spies who had infiltrated American forces. According to a May 1946 article in the *Index-Journal* of Greenwood, South Carolina, he also helped track down a truckload of Gestapo-held gold bullion in Bavaria and received a Bronze Star for "heroism and meritorious service that included apprehending four SS men in the town of Wuerzburg while the town was under enemy shellfire and heavy street fighting was in progress." The article noted that "valuable information of a counterintelligence nature was obtained in Wuerzburg during his operations."

After the war, Blake returned to the Field Museum and in 1953 published *The Birds of Mexico*, the Mexican equivalent of Bond's *Birds of the West Indies*. He claimed that despite his many expeditions, he never went on an adventure. "A professional doesn't look for adventures. To have an adventure is to admit stupidity and incompetence. We prefer to call close calls 'acts of God,'" he told the *Chicago Tribune* in 1980.

As Blake also told the *Chicago Tribune*: "There's not much similarity between hunting birds and hunting spies, except that each requires careful planning."

Emmet Reid Blake of the Field Museum worked for US Army counterintelligence in North Africa and Europe during WWII. © *The Field Museum*

Jim Bond leaving his corner office at the ANSP. F*ree Library of Philadelphia, Rare Book Department*

CHAPTER 009

Was Jim Bond a Spy?

Could the real James Bond also have been a spy? Probably not, but there are enough intriguing connections that it's fun to speculate.

In his fictional biography of James Bond (*James Bond: The Authorized Biography of 007*), author John Pearson certainly flirted with the idea. Pearson's story has Fleming discovering the existence of an actual secret agent named James Bond, whom he met in the lobby of a Caribbean hotel: "Even in those days, James was engaged in some sort of undercover work."

The fictional biography also posited that Fleming's real purpose in writing the 007 stories was to make his fictional Bond such a comic-book superhero that the Russians would fail to take the real Bond seriously, thereby allowing him to continue his secret work. Other theorists—notably William Kelly in his blog *James Bond Authenticus*—have explored that possibility, combing through Bond's career and Mary's memoirs for clues pointing to a life of espionage. Could it have been possible?

A case could be made from the circumstantial evidence. Bond attended two schools that educated future spies. He went to the West Indies during an era of Nazi intrigue. During WWII, at least six of his contemporaries affiliated with natural-history museums worked for OSS, and a seventh worked for US Army Counterintelligence.

At several points in Jim Bond's four decades in Cuba, Haiti, and other hotspots in the Caribbean, he would have been a useful source of information for US intelligence agencies.

As Mary Bond wrote in *To James Bond with Love*, her husband was in the Dominican Republic when Rafael Trujillo took power in the early 1930s—and even had the dictator sign his collecting permit. And he was in Cuba in early 1961, shortly before the infamous Bay of Pigs Invasion, in which the United States bankrolled dozens of Cuban expatriates in their efforts to overthrow the iron-fisted Castro regime.

In other instances, Bond certainly was in peculiar places at peculiar times. Ian Fleming himself wrote in *Goldfinger* that something that happened once was happenstance. Twice was coincidence. And three times was enemy action.

Chapter 009

The SS *America*, which had its name and American flags emblazoned on its hull in hopes German U-boats wouldn't torpedo it, was painted in a camouflage pattern when it became the USS *West Point*. *Public domain*

First, his schooling—in theory the perfect academic arc for a future spy. At age twelve, Jim Bond attended St. Paul's School for two school years. The boarding school also educated future spies S. Dillon Ripley and Brooke Dolan II.

The next stop in Jim Bond's academic career was Harrow School in North London. Harrow was also the alma mater of the infamous Richard Meinertzhagen, the notorious British spy and ornithologist. Bond went to the same university as Meinertzhagen, Cambridge's Trinity College—home to Kim Philby and the most notorious spy ring of the twentieth century.

After college and a brief detour as a banker, Bond worked as an ornithologist for the Academy of Natural Sciences, where his contemporaries included Ripley, Dolan, and the other five American operatives described in the previous chapter.

Those connections could lead one to suspect that Bond was somehow involved with the OSS as well, but he would have been in the OSS personnel files if he had worked directly for them, and the agency rarely recruited agents who had gone to college overseas. Besides, it's far more productive to review what is fact, not speculation.

Bond was by all accounts an excellent observer and solid marksman. He could speak limited French and Spanish and by WWII had excellent local knowledge and government contacts in many islands in the West Indies, including Cuba, the Bahamas, Jamaica, Haiti, and the Dominican Republic. His ornithological research in any of these countries would have provided him with perfect credentials to return undercover.

One such story of intrigue involved a trip to Haiti in the late spring of 1941. Europe was already at war, and the United States was seven months away from the Japanese attack on Pearl Harbor. At the time, Haiti was neutral as well. Like the United States, it declared war on Japan the day after their sneak attack had killed more than 2,400 Americans and wounded another 1,178. Haiti declared war on Germany and Italy shortly afterward.

Mary recounted the incident in her 1980 memoir, *To James Bond with Love*:

> If Ian Fleming had lived longer, it's a safe guess that he and Jim (Bond) would have met again. Fiction writers are scavengers when seeking material for their fabrications, and

Was Jim Bond a Spy?

Bond's name was the first one on the passenger list for the SS *America*'s cruise to the West Indies that sailed from New York in May 1941. *National Archives, public domain*

Fleming might have easily extracted from Jim his enigmatic experience in World War II in Haiti. He arrived in Port-au-Prince in May 1941 for the sole purpose of studying birds on Morne La Selle, a plateau over 6,000 feet high. He put up at the little Hotel de Reix at Kenscoff, a small settlement at about 4,000 feet and tried to obtain two porters to carry his camping material. No one would go. A German, he was told by the inhabitants, had built an airstrip high on the ridge and would not allow anyone to go up there. Jim asked where the German lived, and the natives pointed to the summit of nearby Morne Tranchant which is covered by low scrubby woods. Jim was doubtful of this, for he was used to near-enough gestures from the islanders when asked the location of some rare bird, but decided to go up and see for himself.

He climbed to the top of the mountain and found on the edge of a small clearing a very neat cottage well hidden in the foliage. The German who came out was very pleasant, spoke excellent English, and although he did not say, "Dr. Livingstone, I presume," to Jim's astonishment he knew who he was and told him to go ahead wherever he pleased on the La Selle ridge. Jim was so absorbed in his own objectives he forgot all about the alleged airstrip and went on his way without even asking about it.

When leaving Haiti for home, he was forced, owing to the war, to travel on a freighter from St. Marc [in Haiti] to New York. Back in Philadelphia he told his friend Brandon Barringer about the encounter with the German, and Brandon took it up with the authorities in Washington. Jim was promptly visited at the Academy of Natural Sciences by Army, and then Navy, Intelligence officers. Fleming would have been intrigued with the final twist to the story. The Intelligence people asked a lot of foolish questions and seemed far more suspicious about Jim's reasons for climbing Morne La Selle than about the German's activities.

This trip had a couple of other odd aspects. Perhaps it was just a coincidence, but the trip was not listed in the Academy of Natural Sciences' annual proceedings (only five were listed that year). It was the first year for which no Bond trip had been included. But that's only the half of it.

For starters, on May 10, 1941, Bond left New York City on the 723-foot-long SS *America*, bound for Haiti. It made the round trip between New York and the Caribbean every two weeks This United States Lines ship, which had a cruising speed of more than 23 knots, was billed as "the largest, fastest and most luxurious ship ever built in the United States"—a far cry from the cargo ships and mail boats that Bond typically used. In addition, the ship had been secretly designed so that it could be converted into a troop carrier if war with Germany broke out. The SS *America* was launched on August 31, 1939, the day before Nazi Germany invaded Poland. Although the ship had been intended for luxury transatlantic crossings, her permanent route was soon switched to New York and the Caribbean, where she would travel in neutral waters, farther away from the war in Europe and Hitler's U-boats.

The SS *America* dropped off Bond at Port-Au-Prince on May 16. The trip would be the luxury liner's penultimate cruise for eight years. On May 21, in the South Atlantic, a German U-boat stopped the SS *Robin Moor*, an American cargo ship that proclaimed its wartime neutrality with a large "USA" and American flag painted on its hull. After giving its thirty-eight merchant seamen and eight passengers a few minutes to board four lifeboats, the U-boat torpedoed and shelled the *Robin Moor* until it sank.

A 1941 Brochure for the SS *America*'s Caribbean cruises. *Public domain*

Soon after, on June 1, the US Navy requested that the SS *America* be converted into a fully operational transport ship. Over the next two weeks, the 1,202-passenger liner was refitted to carry some 5,400 troops at a time.

On June 20, in a radio address, President Franklin D. Roosevelt said the sinking of the *Robin Moor* was perpetrated by a Nazi U-boat and called the actions "outrageous and indefensible."

That same day, as Peter Duffy recounts in his book *Double Agent*, German expatriates Erwin Wilhelm Siegler and Franz Joseph Stigler, a butcher and a baker who worked in SS *America*'s kitchen, were secretly arrested on charges of attempting to leave the country without notifying their draft board.

On June 28, 250 agents arrested more German spies. In all, Siegler and Stigler and thirty-one others were charged as part of a massive spy ring. Siegler, the ship's butcher,

Was Jim Bond a Spy?

The Duquesne Spy Ring included three SS *America* crewmen—Hartwig Kleiss, Erwin Wilhelm Siegler, and Franz Joseph Stigler. *Library of Congress, public domain*

was one of the Duquesne Spy Ring's organizers and contacts. He obtained information about the movement of ships and military defense preparations at the Panama Canal. He was sentenced to ten years on espionage charges.

Stigler, the ship's baker and confectioner, had sought to recruit amateur radio operators in the United States to communicate with German radio stations. He had also reported defense preparations in the Canal Zone and had advised other German agents. Stigler was sentenced to serve sixteen years.

A third enemy agent who had worked aboard the SS *America*, Hartwig Richard Kleiss, had passed along information to the Germans, including blueprints of the ship that showed the locations of gun emplacements and how the guns would be brought into position for firing. Kleiss also obtained details on the construction and performance of new speed boats being developed by the US Navy, which he tried to pass along to Germany. He pleaded guilty to espionage and received an eight-year sentence.

Of the thirty-three members of the German spy ring, sixteen pleaded guilty and the others stood trial and were convicted. The Duquesne Spy Ring was considered the largest espionage case that ended in convictions in American history.

If Bond did work for American intelligence operations, however unlikely, the US government would have kept a file on him. Under the Freedom of Information Act, a request was made in late 2016 for all documents and files pertaining to ornithologist James Bond.

The CIA's information and policy coordinator, Michael Lavergne, offered this hazy response: "After conducting a search reasonably calculated to uncover all relevant documents, we did not locate any responsive records that would reveal an openly acknowledged CIA affiliation with the subject.

"To the extent that your request also seeks records that would reveal a classified association between the CIA and the subject, if any exist, we can neither confirm or deny having such records, pursuant to Section 3.6(a) of Executive Order 13526, as amended. If a classified association between the subject and this organization were to exist, records revealing such a relationship would be properly classified and require continued safeguards against unauthorized disclosure."

Free Library of Philadelphia, Rare Book Department

CHAPTER 010

Bond's Legacy

The real James Bond probably would have cringed at this book. As his ANSP colleague Robert McCracken Peck commented, "He went under the radar as much as he could and resented any kind of publicity." However, he didn't fly under the radar nearly as much as he would have hoped—thanks in part to his wife and main promoter, Mary.

No biography of Jim Bond would be complete without an assessment of his legacy. Separating Bond from his unwanted 007 connection is difficult, since it has—albeit inadvertently—raised awareness of Caribbean birds and birding for more than half a century. To evaluate Bond's own achievements, one must examine three aspects of his career: his books (notably *Birds of the West Indies* in all its incarnations), his scientific research (including "Bond's Line"), and his impact on ornithology, birding, and the Caribbean.

George Armistead of Philadelphia, who has led dozens of birding tours, including one called "The West Indies: Birding James Bond's Islands," described the impact of *Birds of the West Indies*. "[It] brought to life birds that before then had only been dreamt of, and the magic of the todies, the lizard-cuckoos, the bullfinches, and some very fancy hummingbirds spilled onto the pages. That book was, for the better part of seven decades, the bible for birders visiting the region."

Armistead says, "I think his book was a landmark achievement that made the region as a whole seem suddenly accessible, and surely inspired a lot of travel and was the stepping stone to further research."

When I visited Orlando Garrido in Havana, the legendary Cuban ornithologist peppered his conversation with references to two pioneering ornithologists he most admired, Cuba's Juan Gundlach and James Bond.

Garrido called Bond's *Birds of the West Indies* "the singular tome for bird studies" and said he thought Bond's "Checklist of Birds of the West Indies" and its many supplements were his biggest contributions to ornithology.

Jamaican environmental researcher Catherine Levy, who wrote about the history of Caribbean ornithology for the journal *Ornitología Neotropical*, says that Bond "definitely spurred interest in the birds of the Caribbean, first from a visitor's point of view in providing a field guide for the first time, and providing information to locals."

Because *Birds of the West Indies* was published in so many editions over so many decades, several generations of birders have felt its impact. Consider this recollection from Frederic Briand, head of the Mediterranean Science Commission, for the *National Geographic*'s blog in 2012. In the late 1970s, as Briand was conducting research at Discovery Bay Marine Laboratory in Jamaica, he decided to go inland and explore the Blue Mountains on a day when the seas were too rough for fieldwork. He ended up in a remote forest field station, looking through binoculars at a huge variety of local birds that ranged from hummingbirds to woodpeckers to parrots.

When Briand asked a forest ranger for help in identifying the birds, the ranger handed him a large field guide—"the very best," according to the ranger.

"That copy had seen better days for sure; it was worn out, some pages were missing," Briand wrote. "But it contained hundreds of drawings and annotations depicting the diverse bird fauna of the Caribbean islands. A pioneering study. The name: *Birds of the West Indies*. The first year of publication: 1936. There were many editions to follow. The author: a certain James Bond, a leading American ornithologist, working at the Academy of Natural Sciences of Philadelphia."

Bond's legacy also includes the Smithsonian Institution's James Bond Fund, earmarked for Caribbean ornithological research. He donated a sizable part of his estate to the Smithsonian and not the Academy of Natural Sciences, as one might have expected, for reasons that are not clear. It might have been because of some perceived slight by the academy, or because of his friendship with Smithsonian directors Dillon Ripley and Alexander Wetmore, who had done work on Hispaniolan birds earlier in the century.

As for Bond's research, he is perhaps best known for the theory he proposed in 1934 that Caribbean birds were most closely related to North American birds—not South American birds, as had previously been thought. It took Bond almost three decades to convince many zoologists that he was right.

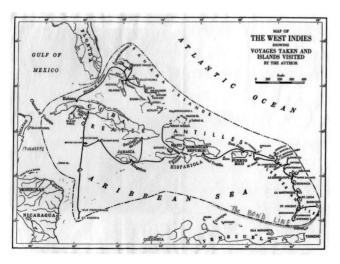

David Lack added "the Bond Line" in red pencil on this map of the West Indies. *Free Library of Philadelphia, Rare Book Department*

"Bond's Line" has since lost some luster as ornithologists have modified it somewhat, but as Keith Thomson, former president of the Academy of Natural Sciences in Philadelphia and former director of Yale's Peabody Museum of Natural History, points out: "It remains a very significant insight."

Bond's other scientific research was far more extensive than most people realize, a fact underscored by the work of zoologist Gerhard Aubrecht of the Austrian Zoological-Botanical Society at the University of Vienna. Aubrecht stumbled across the story of James Bond and Ian Fleming during a trip to Jamaica in 2001 and soon began to collect Bond's scientific papers—which were published in places as far-flung as Cuba, Tobago, and Peru.

Aubrecht then compiled a database for analyzing different aspects of Bond's ornithological work in the West Indies. The database, which took more than a decade to assemble, contains more than 24,000 bird records from the West Indies and adjacent islands that Bond mentioned in his scientific papers. In 2017, the Academy of Natural Sciences published Aubrecht's ten-page "Bibliography of James Bond"—the first ever—which documented 150 papers as well as dozens of Bond's other publications.

"The way Bond meticulously gathered all the information about the Caribbean birds over decades in a kind of private enterprise is astonishing," says Aubrecht. "His analysis of all the data and the results he gained make him an outstanding scientist in island biogeography."

Quantifying the impact of Bond's work is a bit trickier. In an article for *Scientometrics* in 2004, Grant Lewison, senior research fellow at King's College London, assessed the influence of the various editions of Bond's *Birds of the West Indies* by using bibliometrics, a statistical analysis of books, articles, and scientific papers. As Lewison pointed out, it's difficult to assess the value of a book on birds because its readership and citations depend to a large extent on its geographical coverage—a field guide to the birds of North America, for instance, will have a larger impact than one about the Caribbean. Lewison concluded that "*Birds of the West Indies* has proven to be of enduring scientific interest and to have established its author as an ornithologist of distinction."

Nate Rice, ornithology collection manager for the Academy of Natural Sciences, concurs: "Anybody who is studying Caribbean birds is going to cite Bond papers. That's his lasting legacy of scientific importance—how often his papers are cited."

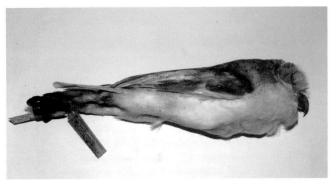

This subspecies of barn owl, *Tyto alba bondi*, was named after Bond. *Photo by Kaylin Martin, courtesy of Carnegie Museum of Natural History*

Bond's first paper, "Nesting of the Harpy Eagle," appeared in *The Auk* in 1927. His last, "Twenty-Seventh Supplement to the Check-List of birds of the West Indies (1956)," appeared sixty years later in 1987, two years before his death.

A world traveler for much of his life, Bond was forced to narrow his range in December 1974, when he had surgery for prostate cancer. The cancer returned, and several years later he developed slowly progressing leukemia. Although Bond would live another fourteen years. Mary wrote in *To James Bond with Love* that there would be "no reading of papers at conferences at London or Basle; no more island-hopping in the West Indies . . . Jim was grounded which presented us with a difficult assignment, but not one that prevented old ties with Caribbean from enduring."

Frank Gill, another noted ANSP colleague, recalls visiting the Bonds during Jim Bond's last years. "He was very engaging, keen, with piercing eyes, and he'd look at you and expect a response. He was not a recluse, but he kept a limited circle. He always wanted to talk about birds and the Caribbean. If it was some other topic, his eyes would glaze over and he'd get back to birds. Like Roger Peterson in that regard, and maybe some of the rest of us."

Bond died peacefully in his bed on Valentine's Day 1989, far from the possible calamities that threatened his famous namesake—being blown up, poisoned, or tortured by a horde of archvillains who are threatening Western civilization. But even in death he could not escape the 007 connection. Reuters' obituary, which ran in *the Washington Post* three days later, characterized Bond as "one of the world's most famous ornithologists" and began (a bit inaccurately, given Bond's fondness for shotguns):

> The actual James Bond, unlike his fictional namesake, never toted a gun and never drank a martini that was shaken, not stirred. He spied on birds, not beautiful female enemy agents, for a living.

James Bond is buried next to his wife. Mary, in the Church of the Messiah Cemetery in Lower Gwynedd Township, Pennsylvania. *Photo by author*

Bond's longtime Academy of Natural Sciences colleague Ruth Patrick summed up Bond's life more on point: "A really great scientist is a person who devotes their life to their science and it is not a person who is here today and gone tomorrow, And Jim was one of those people. . . . His entire life, once he got into birds, he never lost his drive to understand birds."

A month after Bond died, former ANSP president Keith Thomson wrote to his widow, Mary: "He will be missed. I don't know which image is stronger: James as the intrepid traveler to every fascinating corner of the Caribbean—rumpled clothes and uncertain scheduling—or James the precise scientist, immaculately dressed, quiet and thoughtful. Of course, he was all this and much more."

A Few Accolades

Among the awards Bond accumulated over his distinguished career: the Institute of Jamaica's Musgrave Medal in 1952 for his contributions to West Indian ornithology, the Brewster Medal (the highest honor of the American Ornithologists' Union) in 1954, and the Academy of Natural Sciences' Leidy Award in 1975. Only two other Academy scientists have received the award in its long history.

APPENDIX 001

In Bond's Footsteps

The Philadelphia skyline, looking northwest along the Benjamin Franklin Parkway. *Photo by author*

As part of the research for this book, I visited more than two dozen places where Jim Bond lived or worked, including Pennsylvania, Maine, Cuba, and Jamaica. In the process, I discovered three recurrent themes.
- Almost all of the major landmarks in Bond's life still stand today, even if some have changed drastically.
- Bond traveled west of his home state only twice in his entire life, but he sure got around. He went to school in England for eight years. He spent most of his summers in Maine. And he traveled to the Caribbean for much of his life.
- No matter how much he traveled, Philadelphia and the Academy of Natural Sciences were his true home.

Here's a look at James Bond's old haunts today.

Philadelphia and Environs

Jim Bond's birthplace on Pine Street has been divided into several residences. *Photo by author*

1821 Pine Street: The Philadelphia four-story brick house where Bond was born and where the Bond family stayed in the winter in the early 1900s has been divided into apartments.

Academy of Natural Sciences, 1900 Benjamin Franklin Parkway, Philadelphia. Bond worked in this brick building from 1926 to the 1970s. Several photos of Bond—typically smoking a pipe—adorn the walls of the ornithology department. His corner office, which overlooked the Franklin Institute and Logan Circle, is now an ornithology lab.

Birds, insects, and fish that Bond brought back from the West Indies are part of the institution's collection of more than seventeen million plant and animal specimens. Fellow contributors to that collection include John James Audubon, Ernest Hemingway, and Lewis and Clark. The academy, the oldest natural-sciences institution in the Western Hemisphere, acquired a new affiliation in 2011 and is now Academy of Natural Sciences of Drexel University.

The Academy of Natural Sciences of Drexel University, where Bond worked for five decades, is home to his archive and many of the birds, fish, and insects he collected in the West Indies. *Photo by author*

Chateau Crillon Apartment House, 222 South 19th Street, Philadelphia. When Bond lived there before getting married in 1952, the twenty-seven-story high-rise was called the Crillon Hotel. It was designed by Horace Trumbauer, who also designed the Bonds' Willow Brook estate.

Church of the Messiah, 1001 Dekalb Pike, Lower Gwynedd Township. Jim and Mary Bond's ashes are buried in the cemetery near the graves of Jim's mother, brother, and sister. The church's altar has an exquisite stained-glass window donated by Bond's father in memory of Bond's mother. The window, depicting a resurrection scene, was made by Clayton & Bell of London, who also designed several windows for Westminster Abbey.

Bond's grave is marked with a stone marker that says simply, "James Bond, 1900–1989." It's 2 miles from Willow Brook, his childhood home.

Davidson Road, Chestnut Hill, Philadelphia. The house where the couple lived in the 1960s was designed by noted midcentury architect and sculptor Oscar Stonorov, a German émigré who opened a Philadelphia office with fellow architect Louis Kahn.

In her memoir *Ninety Years at Home in Philadelphia*, Mary wrote that the house "was built of soft rosy brick, glass, and aluminum tucked into the side of a ravine. The flat roof was but one foot higher than the street." Bond used to take walks in the nearby Wissahickon Valley Park.

When I stopped by in the spring of 2018, the house appeared to be unoccupied and was in disrepair.

Hill House, 201 W. Evergreen Avenue, Chestnut Hill, Philadelphia. Jim and Mary Bond lived in this eleven-story apartment high-rise by the Chestnut Hill train station from the early 1970s to his death in 1989.

Ingersoll/Clayton House, Old Bethlehem Pike, Spring House, Lower Gwynedd Township. The township now owns the large stone house where Jimmy Bond spent the spring and autumn of 1907 while the family was building a mansion nearby. The house is boarded up, and the grounds are off-limits to the public.

Appendix 001

Bond worked at the Pennsylvania Bank after his graduation from Trinity College; the building is now home to a Del Frisco's Double Eagle Steakhouse. *Photo by author*

Pennsylvania Bank Building, 1426–1428 Chestnut Street, Philadelphia. Bond worked here in the Foreign Exchange Department in his first job out of college in the early 1920s. The bank is now an elegant Del Frisco's Double Eagle Steakhouse complete with the original revolving doors at the front entrance, enormous marble pillars, and a special dining area in the vault on the lower level.

The Drake, 1512 Spruce Street, Philadelphia. Jim and Mary lived in a two-bedroom apartment in the thirty-story art deco apartment building after they got married. The living room was large enough for Mary's grand piano.

Thomas Bond House: Although there's no record of Jim Bond visiting the home of his ancestor, the cofounder of Pennsylvania Hospital, you can visit—and stay overnight. The four-story brick house, at 129 South 2nd Street in Philadelphia's Independence National Historical Park, is now a twelve-room bed-and-breakfast decorated in eighteenth-century federal style.

Willow Brook, 1325 Sumneytown Pike, Lower Gwynedd Township. It's easy to see how young Bond got hooked on nature during the few years he and his family spent their springs and autumns on this 350-acre country estate. The Bonds planted all sorts of bird-friendly shrubs and trees along what is now known as Rhododendron Drive, and the huge property has all sorts of great bird habitats, from wetlands and creeks to fields and woods.

Even on a mid-January walk around the campus, I saw plenty of avian activity, including nearly a dozen eastern bluebirds, several woodpeckers, two turkey vultures, a red-tailed hawk, and a tree full of American crows. The property was sold when Bond's widower father remarried in 1913.

Gwynedd Mercy University's campus has occupied the heart of Willow Brook since 1948. Assumption Hall, the school's enormous brick administration building, was originally the Bonds' mansion. The fully restored building later became a convent, and when the nuns invited Jim and Mary Bond back in the 1970s, they had a reception in the house's formal dining room. Bond told the other attendees that this was the first time he had eaten there—as a child he always had to eat in a dining room set aside for the children.

The university occasionally makes note of the Bond connection, including a video produced by the admissions department a few years ago that features a 007-like Jim Bond searching for the school mascot—a mythic half-bird, half-wildcat known as a griffin.

In Bond's Footsteps

Willow Brook, Jim Bond's childhood home, is now part of Gwynedd Mercy University. *Photo by author*

Mt. Desert Island, Maine

Bond summered here for his entire life, most notably at his uncle Carroll Tyson Jr.'s summer home and nearby hunting camp, and later at a cabin he owned with Mary on Pretty Marsh. The largest island in Maine wasn't much of a birding destination back when Bond and his uncle wrote *Birds of Mt. Desert Island, Acadia National Park, Maine* in 1941, but it has become a popular birding spot and home of the annual Acadia Birding Festival.

Bond's presence on Mt. Desert Island has faded. If you ask at the front desk of the public library in Southwest Harbor or Northeast Harbor, you can see (but not borrow) Bond's groundbreaking booklet—Southwest Harbor even has a hardbound version of the softcover pamphlet. And you can see the same yellow-covered pamphlet in Acadia National Park's archives if you make a reservation in advance.

Acadia Birding Festival, Somesville. The popular four-day event has been held annually since 1999, drawing serious birders from as far away as Arizona. Bond's name is invoked from time to time, but the festival is focused on the present, with dozens of events ranging from guided bird walks (including one to the shores of Pretty Marsh) to talks by big-name guest speakers.

The festival center, in the local firehouse's community room, offers a wide array of modern field guides, brochures for birding trips to such Bond destinations as Cuba and Jamaica, and binoculars so sophisticated they may have changed Bond's mind about using a pair (although the tight-fisted Bond would have needed convincing).

Birders at the Acadia Birding Festival look for warblers each June not far from Jim and Mary Bond's summer residence in Pretty Marsh on Mt. Desert Island, Maine. *Photo by author*

Birchcroft and Stoneledge, Northeast Harbor. These stately cottages were Carroll Tyson's summer home base. Bond spent his summers there from the 1910s to the 1950s. The elegant eight-bedroom Birchcroft is still privately owned and enjoyed by seasonal visitors. Tyson painted his celebrated portfolio of Maine birds at Stoneledge, which is also still privately owned. One of those prints, of blue jays raiding a robin's nest, hangs on a wall on the first floor of Birchcroft.

Pretty Marsh Camp, off Pretty Marsh Road, Pretty Marsh. The cabin and outbuildings where Jim and Mary Bond spent their summers for more than three decades have recently been restored. The property is still privately owned.

Tyson's Hunting Camp, Long Pond. Tyson's large home and property, near the water at the base of a steep hill, are still privately owned.

Bond spent many summers at Birchcroft, his uncle Carroll Sargent Tyson Jr.'s home in Northeast Harbor, Maine. *Photo by author*

Cuba

Cuba figured larger in Jim Bond's life than in Ian Fleming's. It was one of the best Caribbean islands for ornithology, with more than 370 bird species, including the tiny bee hummingbird and twenty-six other endemics—species that can be seen only in Cuba. For Fleming, Cuba was a source for villains, notably Major Gonzalez, a Cuban gangster trying to buy up land in Jamaica after Castro's takeover in Fleming's bird-filled short story, "For Your Eyes Only."

Bond visited the island many times, including just after the revolution. In many ways, the island has changed little since then, for better and worse. Cuba is still full of amazing natural beauty and untapped tourism potential.

Playa Larga on the Bay of Pigs is now a resort town. *Photo by author*

Bay of Pigs. When the Bonds visited in 1960, they scoffed at Castro's plans to build a resort on the Bay of Pigs. These days, many European and Russian tourists head for the beach town of Playa Larga, where you can enjoy the sandy beach and see a spectacular sunset. You can also watch a neighborhood peddler selling bread door to door from a horse-drawn wagon while 1950s Chevrolets and modern European-built taxis zip past—a taste of three centuries all at once.

Havana. Cuba's capital is reminiscent of Jim Bond's Philadelphia of the 1950s in one major way: the same American cars spewing thick fumes.

The Presidente, the hotel where Jim and Mary Bond stayed, is still owned by the government, but the room rates are no longer $5.10 a night—they list for closer to $200.

The Cuban tody, which graced the front of the dust jackets for Bond's 1936 and 1947 editions of *Birds of the West Indies*, is still abundant in the Zapata Swamp. *Photo by author*

Zapata Swamp. Cuba is rightly proud of the Zapata Swamp and the 2,425-square-mile Ciénaga de Zapata Biosphere Reserve, considered by UNESCO to be one of the largest and most important wetlands in the Caribbean region. But back in the days before Cuban expatriates funded by the United States invaded in 1961, Fidel Castro wanted to clear the swamp and use the land to grow sugar cane. Fortunately, the Zapata Swamp remains wilderness.

On a visit in November 2016 with the Caribbean Conservation Trust, a nonprofit group that does regular bird surveys in Cuba, I visited the swamp in hope of seeing two of the swamp's most famous residents, the Zapata wren and the Zapata sparrow. A third celebrated resident, the Zapata rail, had rarely been seen or heard in recent years. All three were Bond's target birds (literally and figuratively) back in the late 1920s.

Helping lead the way was Frank Medina, the Cuban official who runs the reserve. A tourist from Canada had given Medina her copy of Bond's *Birds of the West Indies* on a visit in the early 1980s, and that was how he learned about his nation's birds.

While we were looking for the wren and sparrow—successfully, it turned out—a Cuban tody appeared, seemingly from out of thin air. The Zapata wren and Zapata sparrow may be far rarer, but the Cuban tody is the showstopper, a miniature living, breathing rainbow of bright pastels. When Mother Nature was handing out beauty to the bird kingdom, the Cuban tody came back for a second helping. No wonder Bond put it on the cover of the 1936 and 1947 editions of *Birds of the West Indies* and featured it on the full-color frontispiece as well.

Jamaica

To understand Jim Bond or Ian Fleming, you had best visit Jamaica. For Fleming, the island was both sanctuary and workplace. Not only did he invent his fictional James Bond and write all of the 007 novels in Jamaica at Goldeneye, but the island nation also provided the setting for *Dr. No*, *Live and Let Die*, and *The Man with the Golden Gun*, as well as the short story "For Your Eyes Only," which starred red-billed streamertails. This is where Fleming was happiest.

For Bond, Jamaica was one of the key islands of research for *Birds of the West Indies*. The attraction was clear. Jamaica was, and is, known as a biodiversity hotspot where you can find birds you'll likely see nowhere else. It ranks fifth among the world's islands for endemic species. By one count, the island has twenty-nine endemics. The island nation, which remained a British colony until 1962, is also the place Bond and his wife kept returning to long after his research days were over, for the same reason that two of America's finest nineteenth-century artists, Frederic Edwin Church and Martin Johnson Heade, kept returning: the island is flat-out beautiful.

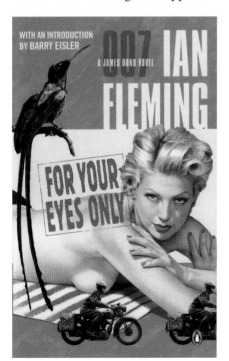

This 007 paperback featured a streamertail, the national bird of Jamaica and a Fleming favorite. Front cover in its entirety from *For Your Eyes Only* by Ian Fleming (Penguin Books, 2006). Copyright © Penguin Books Ltd., 2006

Appendix 001

What is known of Bond's travels in Jamaica is largely the result of Mary's memoirs, and it's clear that two Jim Bonds visited Jamaica, the youthful explorer who roughed it in the 1930s, and the older ornithologist of the 1960s and early 1970s. The younger Bond traveled by foot or mule, ate with local farmers, and slept in hammocks. The older Bond and his wife traveled in style by car, sipping cocktails and staying at hotels that were once the great houses of Jamaica's old sugar plantations.

These stately homes and their vast estates were also favorite haunts of Ian Fleming, who stayed at his buddy Ivar Bryce's great house, Bellevue, on his first visit to Jamaica in 1943—for a conference on combating the Nazi U-boats that were wreaking havoc on shipping in the Caribbean.

For the boastful Fleming, the great houses also provided a challenge. According to biographer John Pearson, Fleming once wrote that he was determined that Goldeneye would become better known than any of the great houses "that had been there so long and achieved nothing." One can argue about the great houses' historical value, but the past and present Goldeneye is the best known.

Blue Mountains. In eastern Jamaica, these mountains and their smaller cousins, the John Crow Mountains, dominate the landscape. They are not only steep, with a top elevation of 7,400 feet, but they are also on Kingston's doorstep.

How the Blue Mountains got their name is unclear, although many believe it's because of the blue-gray clouds that often shroud parts of the mountains. Up close, the terrain is several shades of green, as lush and vibrant as any tropical rainforest. Fleming, a world traveler, described the Blue Mountains as the most intoxicating landscape he'd ever seen.

These mountains are also where both Bond and Fleming went to find rare birds. And that wasn't easy back in the day. According to Matthew Parker in *Goldeneye: Where Bond Was Born*, Fleming described being dragged somewhat unwillingly from "the sunbaked sand of my pirate's cove" to ride donkeys up the mountain. In the 1930s Bond likely traveled up the mountain by horse, mule, or donkey as well.

Ramsey Acosta, one of Fleming's gardeners at Goldeneye, says Jamaica's roads back in the thirties "were very bad. There were asphalt roads only along the coasts where the tourists traveled." Even today, the narrow, two-lane roads that thread their way over the Blue Mountains are a challenge to navigate, but for birders in search of mountain witches, Jamaican owls and rufous-crested solitaires (the bird with the haunting call for which Ian Fleming named his femme fatale in *Live and Let Die*), this is heaven.

Ramsey Acosta was Fleming's gardener at Goldeneye for many years. *Photo by author*

A view of the GoldenEye Resort in Jamaica. *Photo by author*

Exterior view of Fleming's Jamaica home, now a coveted part of the GoldenEye Resort. *Mark Collins photo, courtesy of Island Outpost*

Appendix 001

GoldenEye. On the North Coast Highway, not far from the Ian Fleming International Airport and James Bond Beach, sits Ian Fleming's former getaway. When Jim and Mary visited, she described it as a small property tucked away on an isolated promontory outside the little town of Oracabessa: "We found the white gates hospitably ajar between two pink plaster gate posts," she wrote in *To James Bond with Love*. "The gravel driveway dwindled to a mossy track that ended at a low rambling house where, in true West Indian fashion, a long hibiscus hedge was strewn with towels and napkins spread out to dry."

Fleming's friend and neighbor, Noel Coward, described it this way to Fleming biographer John Pearson: "Goldeneye, of course, is a positively ghastly house. I should know. Ian lent it to me for three months the year after he'd built it (and charged me fifty pounds a week for the privilege, I might add, and as I told him, that was too much for bed and board in a barracks). There was no hot water in those days. Only cold showers. We were very manly and pretended to like it."

Coward liked to joke that the austerely modern bungalow looked so much like a medical clinic that it should be called "Golden Eye, Nose and Throat."

Although criminals are notorious for returning to the scene of the crime, it turns out that in the case of Fleming's identity theft, the victim returned as well. On their final trip to Jamaica in the early 1970s, Jim and Mary Bond stopped by to give Fleming's housekeeper, Violet Cummings, a copy of Mary's *How 007 Got His Name*. Mary later recounted in *To James Bond with Love* that although much about the place remained unchanged, down to the towels drying on the hibiscus, she concluded that "the fun was over without Fleming around, and we drove away."

But times change, and Goldeneye's "white gates hospitably ajar" have given way to a high metal fence and locked black iron gates and a guard at the sentry post just beyond the entrance—befitting its current status as an ultra-exclusive seaside resort. The driveway, no longer mossy, splits just beyond the entrance. To the right is the private driveway to Fleming's original home, now called the Fleming Villa. To the left is the main thoroughfare leading to the main office and the many luxury villas, cottages, and huts, most of which are on their own separate island. It is as close to a tropical paradise as one is likely to encounter anywhere.

A mango tree donated by 007 actor Pierce Brosnan is part of the resort. *Photo by author*

In Bond's Footsteps

The beach where key scenes of the movie *Dr. No* were filmed is now called, appropriately, James Bond Beach—complete with restaurant, bar, and Wi-Fi. *Photo by author*

In true Jim Bond fashion, I tried to do some birding while I toured the grounds, with an eye out for the Jamaican grackles that had upstaged Fleming during his Canadian Broadcasting Corporation interview more than a half century earlier. At the resort's waterside Bar Bizot, I found several of them, still making their distinctive two-note "kling, kling" call, and still up to their usual tricks. Two were hopping about on a table freshly set with silverware, plates, and water glasses, with one of them pulling on a white linen napkin apparently just for the fun of it.

While paying due tribute to its previous owner, down to copies of many of the 007 books in each villa, cottage, and luxurious hut, the resort is as much about the movie 007 as the literary 007, and that's as it should be. The cinematic James Bond can reinvent himself every few years, whereas Fleming's literary 007 is frozen in time with both his flaws and strengths. It's significant that the Ian Fleming Room, just a few yards from the old gazebo, features a huge black-and-white photo of Fleming and the original cinematic 007, Sean Connery. It was no doubt taken at the old Goldeneye or the nearby stretch of sand now known as James Bond Beach. To choose a photograph that features Connery so prominently is perhaps an acknowledgment that visitors might need a little visual cue to recognize that Ian Fleming is in the photo as well. Both author and resort can bask in Connery's radiance. Ultimately, the fictional 007 has upstaged his creator in much the way 007 upstaged Jim Bond so many years ago. Fleming's most iconic secret agent remains larger than life, while Fleming and Bond were mere mortals.

GoldenEye is a reflection of its owner, Chris Blackwell, who knows how to pamper the superrich, but it is also a projection of what the old Goldeneye might have become if Fleming had had more time and money. Although Fleming had built the original Goldeneye in a spartan manner, he always admired both luxury and sophistication and made them 007 staples. Today's GoldenEye would suit him just fine.

Mona Great House, on Great House Boulevard, a mile from the University of the West Indies campus. It was from here, on February 5, 1964, that Jim and Mary hired a driver to take them on their surprise visit to Goldeneye.

In *To James Bond with Love*, Mary described the Mona Hotel as such:

> This . . . was another beautiful Colonial Great House built on a knoll four miles outside of Kingston, standing like an elegant grande dame aware of her dignified past. Beyond the terraced gardens the fabulous Blue Mountain range, fold upon fold, caught the ever-changing light of day, composing patterns of color as grand as the greatest symphonies. Two ancient and magnificent mango trees, one on each side of the house, provided shade for tables and chairs, which made a rum swizzle with a sandwich lunch or evening cocktails moments to be remembered.

Mary also lamented that "the Mona Great House was being surrounded with a townhouse development. No use wringing our hands. We could only wonder when the lovely house with its vine-covered walls and white pillars would, itself, be demolished."

She would be pleased to know the great house still stands, and so do the mango trees on either side of it. As she feared, housing obstructs the view of the mountains, and the road through the development runs right next to the great house, depriving it of its front lawn and its former spectacular setting.

The Mona Hotel, a favorite of Jim and Mary Bond's when they visited Jamaica, is now a private residence. *Photo by author*

APPENDIX 002

Goldfincher
by Avian Flemish

Appendix 002

Illustration by Anna Raff

In 1964, The Auklet, an "Irregular Journal of Irreverence" published by American Ornithologists' Union (now the American Ornithological Society), featured this 007 parody, "On Her Majesty's Ornithological Service." It is reprinted with permission, lightly edited.

The first hint of the nectarine-hued Caribbean dawn slipped between the shutters and dappled the rumpled silken sheets on the oversized bed. As a Bare-eyed Thrush uttered its first tentative notes in the lush garden of the hotel, a somewhat more extensively bare figure stirred restlessly and then leaped up.

"What is it, James?" murmured a sleepy voice from deep within the pillow.

"Sorry my dear, but I have already overslept. Who knows what evil schemes may have already been set into motion by that arch-villain and enemy of Her Majesty's Avifauna, the notorious Goldfincher?"

With this the lean, saturnine James Blond hastily zipped himself into his sharply creased field clothes, and opened a secret compartment in the bedpost. From this he extracted the silver .410, the badge of that little band of Her Majesty's most trusted servants who are Licensed to Collect. Ignoring the sultry entreaties to return to bed, Blond silently let himself out of the French windows.

Meanwhile half-way across the island, the same impartial dawn light glinted from the oily nose of a pink-faced obese white man, of uncertain European extraction, who sat with his three servants around a greasy table in front of a native bar.

"It is not necessary," he said coldly, "for you to know all the details. You are being well paid for your part in this project." The tallest of the three servants, a rodent-faced alcoholic named Snitch, whined uncomfortably. "But sar, it is the danger to ourselves we fear. What will these strange glowing seeds do to our manhood?"

Goldfincher sighed ponderously, setting up a minor seiche in his nether dewlap. "All right then, I will explain the whole project, although your feeble minds can never comprehend the glory of it. Here in the Lesser Antilles there are hatfuls of very rare birds. Endemics, they are called. They are very, very valuable and several disreputable Museum Curators have offered me respectable sums of money to get them each a Statistically Significant Series. So I am cornering the market in Lesser Antillean Endemics."

"But sar," said the second servant, a small, grimy dope-addict named Bletch. "Even we, master hunters that we are, cannot be sure that we will get all of these birds. If we miss any perhaps somebody else will get them and offer them on the black market at a lower price."

Goldfincher chuckled, the sound sending a nearby mongoose screaming into the undergrowth. "That is the beauty of my scheme. When you have captured as many specimens of these Endemics as you can, you will blanket the island with my specially prepared Radioactive Birdseed. In this way any remaining individuals will be rendered sterile and the populations will inevitably disappear. Even James Blond will be thrown off the track when my elite corps of rumour-mongers places the blame on a harmless household detergent. And I, Goldfincher, will have the only available fresh specimens of Lesser Antillean Endemics!"

Once again he chuckled evilly, causing a passing Ruddy Quail Dove to blanch in terror.

Several days later, as Goldfincher fretted at the slowness with which his contraband Series of Endemics approached Statistical Significance, James Blond lounged under a chestnut-sided beach umbrella in front of the hotel and sipped at a Pimm's Cup while idly identifying immature terns. A shadow fell across his bronzed torso, and he looked to his right to see a lovely red-headed girl of some 22 years and some 96 centimetres pectoral circumference, clad principally in an abundant coat of freckles to which had fetchingly been added a bikini in the Black Watch tartan.

"Mr. Blond?" she half-whispered.

Blond eyed her tentatively. "And if I were he?"

"Oh Mr. Blond, you are the only one who can help me! I am Angelina MacDreft, only daughter of Angus MacDreft, manufacturer of a harmless household detergent. Some unspeakable persons have been spreading rumours that my father's product has been responsible for a noticeable reduction in the Roadside Counts of Territorial Males of Lesser Antillean Endemics, and if these rumours are not stopped he will be ruined!"

She paused to inhale deeply, an act which Blond watched appreciatively, having been in his younger days a student of the populations of *Parus major* in the Hie'lands.

"Can you help me?" she pleaded. "I'll do anything."

"No doubt, no doubt," said Blond, who had read several of these stories before. "This has all the field marks of Goldfincher's work."

The girl, somewhat confused, looked about her but saw only a small flock of Yellow Grass Finches. She eyed Blond narrowly, hoping she had not placed her faith in a man who could not tell *Sicalis* from *Spinus*, but he went on to explain that he had been on the trail of the evil Goldfincher ever since he had introduced Giant Cowbirds into the Kirtland's Warbler Refuge.

"I think I can help your father, my dear," said Blond, "but first we must have some figures for comparison."

The girl was not surprised at this, having heard of Blond's insatiable appetite for foldouts-of-the-month, but soon realised he referred to Breeding Bird Censuses.

"Come with me," he said, tossing aside his custom Armee-Ruckstande 8 × 40 Featherweights. Holding Angelina's hand, he loped to the hotel garage, where the head mechanic quickly readied for active service Blond's motorcycle. After gallantly ushering the redhead into the sidecar, Blond leaped into the saddle and they were off in a roar of exhaust and a spray of gravel.

Twenty minutes later, having covered every road and trail on the island, they were back at the hotel, Angelina clutching the notebook into which she had been scribbling numbers as fast as Blond called them out.

"It's true," said Blond, glancing quickly through the notebook. "The population of Lesser Antillean Endemics is down by no less than 12.47 percent. Clearly something is afoot."

"Yes," said Angelina, "it is mine. You have parked your motorcycle on it."

"Forgive me, my dear," said Blond, moving the cycle, "I was preoccupied. We must find Goldfincher before it is too late and the Lesser Antillean Endemics join the lost legions of the Passenger Pigeon and the Three-Dollar Annual Dues."

"But how, James? Goldfincher and his evil henchmen might be anywhere on the island and no doubt are well aware of your mission to destroy them."

"You, my dear, must act as bait. We will disguise you as a Lesser Antillean Endemic in breeding condition and they will without question follow you to your nest."

Blond flung open his kit-bag and extracted two lengths of cloth. One, which was Dark Brown to Olive Grey (becoming Dusky anteriorly), he fashioned from the top of her head to her shapely pygostyle. The other, which was Buffy Greyish Brown to Greyish White, he draped about her front, leaving a large oval opening in the middle through which her deliciously freckled abdomen could pass muster in the dim light of the rainforest as an incubation patch.

Whipping from its sheath his gleaming Pfadfinder Special, Blond quickly whittled a long, slender, curved bill which he fastened on her nose. "Now, my pet, let me see you tremble."

A convulsive series of shudders wracked the girl's disguised torso.

"Perfect!" cried Blond exultantly, "Cinclocerthia to the life!"

He helped Angelina into the sidecar once more and they roared off to the nearest remaining patch of rainforest. Blond concealed the cycle under the rubble of the forest floor and with extreme care placed three greenish blue eggs at the base of a palm frond. He then climbed into the dense foliage of a tree fern some 20 metres from the eggs and called in a harsh whisper, "All right, my dear, remember what I have taught you!"

Angelina hopped cautiously to the edge of the patch of rainforest about half a kilometre from where Blond was concealed. She bounded into the lower branches of a tree and began calling a loud, somewhat tremulous "ture-ture-ture-ture," trembling the while.

Not 10 minutes later a sinister black touring car slammed to a halt on the edge of the rainforest.

"Look, you fools!" whispered Goldfincher. "It's a Cinclocerthia—or Grive Trembleuse in your local patois. I thought you had them all!"

The three ruffians and their leader got out of the car and began to walk quietly into shotgun range. Snitch raised his double-barrel to his shoulder.

"Stop, you imbecile!" hissed Goldfincher, pushing the gun down, for Angelina, obeying instructions to the letter, was now hopping along the ground dragging her left arm in an unmistakable Distraction Display.

"There is certainly a nest nearby," said Goldfincher, "and a certain Curator who shall be nameless has offered a handsome bonus for a clutch of Trembler eggs. Spread out!"

Unerringly, Angelina led the four villains to the glade where Blond perched in concealment and the three greenish blue eggs awaited their fate.

"There they are!" shrieked Goldfincher, stumbling forward in greedy haste. "Never mind the bird, the eggs are far more valuable!"

The bloated Goldfincher and the three servants gathered about the nest, and just as Goldfincher was reaching into his collector's bag for his blowpipe, Blond withdrew from his pocket a device which he had modified that morning from a TV armchair channel changer.

With a quiet smile of satisfaction he pressed the scarlet button. A tremendous explosion rocked the rainforest as the three mock trembler eggs responded to the radio waves emitted from Blond's tiny transmitter.

Blond slid down the tree fern, walked over to the shattered remains of the four would-be egg-poachers and looked down at them broodingly.

"So perish all enemies of Her Majesty's Endemic Avifauna!" he murmured.

Then a faint whimper at the edge of the clearing caught his ear. "My God!" he said, "it's Angelina!"

The girl, still wrapped in her disguise, lay crumpled on the ground and Blond could see that her trembling was now no counterfeit. He turned her over gently only to find that a flying piece of eggshell had penetrated her fair bosom with soon-to-be-fatal results.

"Oh James!" she whispered, "now my father's name can be cleared, but I wish that Fate had been kinder to us!"

"I know, my dear," said Blond cradling her long-billed head in his lap. "But you are dying in a noble cause, in Her Majesty's name, and your death will not be futile. I know a museum that will give me fifty pounds for the skin of an adult Cinclocerthia in breeding condition."

And as Angelina's eyes glazed, Blond took out his silver pencil and began to enter her data in his field catalogue.

Bibliography

Bibliography

Interviews (by author, unless otherwise noted)

Ramsey Acosta, March 2017
George Armistead, May 2017
Harry Armistead, October 2016
Gerhard Aubrecht, May 2017
James Bond (interview with David Contosta, October 1985, Chestnut Hill College archives)
Mary Bond (BBC interview, 1966, in the Free Library of Philadelphia archives)
David Contosta, May 2017
Hiroki Fukuda, May 2017
Orlando Garrido, November 2016
Frank Gill, February 2016
Caitlin Goodman, May 2017
David Levesque, June 2017
Catherine Levy, March 2017
Rich MacDonald, July 2017
Ricardo Miller, March 2017
Ruth Patrick (interview with David Contosta, November 1991, Chestnut Hill College archives)
Robert McCracken Peck, January 2016
Herbert Raffaele, May 2017
Nate Rice, July 2016
Mark Ridgway, March 2018
David Allen Sibley, October 2016
Jonathan Smith, January 2017
Keith Thomson, January 2018
Samuel Turvey, July 2016
Charles Tyson Jr., October 2017

Archives

Academy of Natural Sciences of Drexel University, Library and Archives, James Bond collection.
Chestnut Hill College, David Contosta's James and Mary Bond collection.
Free Library of Philadelphia, Rare Book Department, Mary Bond collection.
National Archives (College Park), OSS files pertaining to W. Rudyerd Boulton, James Paul Chapin, Frederick E. Crockett, Herbert Girton Deignan, Brooke Dolan II, and S. Dillon Ripley II.
University of Indiana, Lilly Library, Ian Fleming collection.
University of South Carolina, South Caroliniana Library (Emmet Reid Blake papers).

Auction Catalogs

Profiles in History, catalog for Hollywood Auction 33 (movie memorabilia), Calabasas Hills, California, December 11, 2008. Lot 103.
Sotheby's, catalog for auction of English Literature and History at Aeolian Hall in London on July 11, 1996. Lot 283.

Books

Allen, Joel Asaph. *Biographical Memoir of Elliott Coues 1842–1899*. Washington, DC: National Academy of Sciences, 1909.
Bailey, Martin. *The Sunflowers Are Mine*. London: Frances Lincoln, 2013.
Baltzell, E. Digby. *Philadelphia Gentlemen*. Chicago: Quadrangle Books, 1971.
Barbour, Thomas. *Cuban Ornithology*. Cambridge, MA: Nuttall Ornithological Club, 1943.
Barrow, Mark V., Jr. *A Passion for Birds*. Princeton, NJ: Princeton University Press, 1998.
Blake, Emmet R. *Preserving Birds for Study*. Chicago: Chicago Natural History Museum, 1949.
Bond, Mary Wickham. *Far Afield in the Caribbean*. Wynnewood, PA: Livingston, 1971.
———. *How 007 Got His Name*. London: Collins, 1966.
———. *My Cove Has Many Moods*. Philadelphia: Almo, 1990.
———. *Ninety Years "at Home" in Philadelphia*. Bryn Mawr, PA: Dorrance, 1988.
———. *To James Bond with Love*. Lititz, PA: Sutter House, 1980.
Boulton, Rudyerd. *Traveling with the Birds*. Chicago: M. A. Donohue, 1933.
Bryce, Ivar. *You Only Live Once*. London: Weidenfeld & Nicolson, 1974.
Capstick, Peter Hathaway. *Warrior: The Legend of Colonel Richard Meinertzhagen*. New York: St. Martin's, 1998.
Cargill, Morris. *Ian Fleming Introduces Jamaica*. London: Andre Deutsch, 1965.
Carriker, Melbourne R. *The Bird Call of Rio Beni*. Santa Barbara, CA: Narrative Press, 2005.
Cassin, John. *Illustrations of the Birds of California, Texas, Oregon, British and Russian America*. Philadelphia: Lippincott, 1856.
Chansigaud, Valerie. *All About Birds*. Princeton, NJ: Princeton University Press, 2010.

Bibliography

Contosta, David. *The Private Life of James Bond*. Lititz, PA: Sutter House, 1993.
Corera, Gordon. *Operation Columba*. New York: HarperCollins, 2018.
Coues, Elliot. *Field Ornithology*. Salem, MA: Naturalists' Agency, 1874.
Craven, Wesley Frank, and James Lea Cate, eds. *The Army Air Forces in World War II*. Vol. 7, *Services around the World*. Chicago: University of Chicago Press, 1958.
Craven, Wesley Frank, and James Lea Cate, eds. *The Army Air Forces in World War II*. Vol. 7, *Services around the World*. Washington, DC: Office of Air Force History, 1983.
Duffy, Peter. *Double Agent*. New York: Scribner, 2014.
Fleming, Ian. *Dr. No*. London: Jonathan Cape, 1958.
———. *For Your Eyes Only*. London: Jonathan Cape, 1960.
———. *Live and Let Die*. London: Jonathan Cape, 1954.
———. *Man with the Golden Gun*. London: Jonathan Cape, 1965.
———. *Thunderball*. London: Jonathan Cape, 1961.
———. *You Only Live Twice*. London: Jonathan Cape, 1964.
Garfield, Brian. *The Meinertzhagen Mystery*. Washington, DC: Potomac Books, 2007.
Garrido, Orlando, and Arturo Kirkconnell. *Birds of Cuba*. Ithaca, NY: Comstock, 2000.
Helms, Richard. *A Look over My Shoulder*. New York: Random House, 2002.
Jeffrey-Smith, May. *Bird-Watching in Jamaica*. Kingston, Jamaica: Pioneer, 1956.
Knight, Maxwell. *Animals and Ourselves*. London: Hodder and Stoughton, 1962.
Lack, David. *Island Biology*. Berkeley: University of California Press, 1976.
Lewis, Daniel. *The Feathery Tribe*. New Haven, CT: Yale University Press, 2012
Long, Ralph H. *Native Birds Of Mount Desert Island and Acadia National Park*. 3rd rev. ed. Southwest Harbor, ME: Beech Hill, 1982.
Lycett, Andrew. *Ian Fleming*. Nashville: Turner, 1996.
McConnell, Scott. *Witmer Stone: The Fascination of Nature*. witmerstone.com, 2014.
Parker, Matthew. *Goldeneye: Where Bond Was Born*. London: Hutchinson, 2014.
Pearson, John. *James Bond, the Authorized Biography*. London: Century, 2006.
———. *The Life of Ian Fleming*. London: Jonathan Cape, 1966.
Peck, Robert McCracken, and Patricia Tyson Stroud. *A Glorious Enterprise: The Academy of Natural Sciences of Philadelphia and the Making of American Science*. Philadelphia: University of Pennsylvania Press, 2012.
Ridgway, Robert. *Directions for Collecting Birds*. Washington, DC: US Government Printing Office, 1891.
Ripley, S. Dillon. *Trail of the Money Bird*. London: Longman Green, 1947.
Rosenbaum, Marion K. *Gwynedd Mercy College*. Charleston, SC: Arcadia, 2006.
St. Paul's School Admission Register, 1912–13.
Staff of the Smithsonian Institution. *A Field Collector's Manual in Natural History*. Washington, DC: Smithsonian Institution, 1944.
US War Department. *War Report: Office of Strategic Services; Operations in the Field*. Volume II. Washington, DC: US Government Printing Office, 1949.
Weidensaul, Scott. *Of a Feather: A Brief History of American Birding*. New York: Harcourt, 2007.

Williams, Susan. *Spies in the Congo: The Race for the Ore That Built the Atomic Bomb.* London: Hurst, 2016.

Author's note: Detailed information on the various editions of *Birds of the West Indies* and *Birds of Mt. Desert Island, Acadia National Park, Maine* is provided in chapter 007.

Magazines

Aubrecht, Gerhard. "Bibliography of James Bond (1900–1989)—American Ornithologist—with New Taxa Described." *Proceedings of the Academy of Natural Sciences of Philadelphia* 165, no. 1 (September 2017): 81–91.

Fergusson, James, ed. Special Issue: Ian Fleming & Book Collecting. *Book Collector* 66, no. 1 (Spring 2017).

Ferree, Barr. "Willow Brook House." *American Homes and Gardens*, October 1909.

Fleming, Ann (writing as Mrs. Ian Fleming). "How James Bond Destroyed My Husband." *Ladies' Home Journal*, October 1966.

Hellman, Geoffrey T. "Bond's Creator." *New Yorker*, April 21, 1962.

———. "Curator Getting Around." *New Yorker*, August 26, 1950.

Lewison, Grant. "James Bond and Citations to His Books." *Scientometrics* 59, no. 3 (2004): 311–20.

Nolan, William F. "Ian Fleming." *Rogue*, February 1961.

Olson, Storrs L. "Correspondence Bearing on the History of Ornithologist M. A. Carriker Jr. and the Use of Arsenic in Preparation of Museum Specimens." *Archives of Natural History* 34, no. 2 (2007): 346–51.

Parkes, Kenneth. "In Memoriam, James Bond." *The Auk*, October 1989.

Parkes, Kenneth (writing as Avian Flemish). "On Her Majesty's Ornithological Service." *Auklet*, September 1964.

Peck, Robert McCracken. "To the Ends of the Earth for Science: Research Expeditions of the Academy of Natural Sciences; The First 150 Years, 1812–1962." *Proceedings of the Academy of Natural Sciences of Philadelphia* 150 (April 14, 2000): 15–46.

People column. *Sports Illustrated*, May 24, 1965.

Rust, William J. "Operation ICEBERG: Transitioning into CIA; The Strategic Services Unit in Indonesia." *Studies in Intelligence* 60, no. 1, Extracts (March 2016).

Schuman, Aaron. "*Time* Picks the Best Photobooks of 2013." *Time*, November 24, 2013.

Sidey, Hugh. "The President's Voracious Reading Habits." *Life*, March 17, 1961.

Wylie, Craig. "Francis Beach White." *Alumni Horae* (St. Paul's alumni magazine) 28, no. 1 (Spring 1948).

Newspapers

Apt, Jay. "Rare Birds Hunter Also Meets Snakes." *Philadelphia Daily News*, May 13, 1955.

"Baseball Evolution Wrought by Smart Set." *Washington Times*, March 2, 1903.

Conniff, Richard. "Species Seekers and Spies." *New York Times*, February 20, 2011.

De Burton, Simon. "Buying Ian Fleming's Books for Investment." *Financial Times*, October 4, 2008.

"Emmet Blake Arrives Here Last Night for Visit after 39 Months' Duty in Europe." *Index-Journal* (Greenwood, SC), May 22, 1946.

"Expert Gets Bird New to Science." Likely from *Philadelphia Inquirer*, 1931, in Bond archives at Free Library of Philadelphia Rare Book Department.

"Father Raced to Dying Daughter . . ." *Philadelphia Inquirer*, September 8, 1904.

Frantz, Douglas. "Ornithologist Says Jungle Trips Are for the Birds." *Chicago Tribune*, November 6, 1980.

"Girard's Talk of the Day." *Philadelphia Inquirer*, March 30, 1923.

"Horned Fowl Like Streamlined Turk Found in Bolivia." *Escanaba Daily Press* (Escanaba, MI), October 10, 1939.

Huxley, John. "Long Lenses, Short Skirts and a View to a Thrill." *Sydney Morning Herald*, August 25, 2001.

"James Bond Finds Clue to Curlew Killer," "James Bond in Case of the Vanishing Curlew," "This James Bond Catches Birds Instead of Villains," and other headlines on wire story distributed by the Associated Press, May 13, 1965.

"James Bond, Ornithologist, Spy's Namesake, Dies at 89." *Washington Post*, February 17, 1989.

Kilcullen, Roy. "Buying Ian Fleming's Books for Investment." *Financial Times*, October 4, 2008.

Macdonald, Helen. "Spies in the Sky." *The Guardian*, May 12, 2015.

Macintyre, Ben. "He Dreamt Up Bond, but Did Fleming Also Create the CIA?" *The Times*, August 15, 2008.

Margaret Bond death notice. *Philadelphia Inquirer*, April 29, 1912.

Martin, Pete. "I Call on James Bond." *Sunday Bulletin Magazine* (Philadelphia), October 4, 1964.

Molotsky, Irvin. "He Took Smithsonian out of the Attic." *New York Times*, March 3, 1984.

News item on sale of Philadelphia Phillies. *Pittsburgh Press*, March 4, 1903.

"Notes of the Street." *Philadelphia Inquirer*, December 3, 1909.

"Wrong Man." *Sunday Times* (London), Fall 1960.

Correspondence

Blake, Emmet Reid, to his mother, February 1927, University of South Carolina, South Carolina Library.

Bond, Mary, to Ian Fleming, February 1, 1961, Free Library of Philadelphia, Rare Book Department archives.

Bray, Hillary, to James and Mary Bond, 1964, Free Library of Philadelphia, Rare Book Department archives.

Cadwalader, Charles, to Witmer Stone, August 11, 1931, Academy of Natural Sciences of Philadelphia.

Cornwell, David, to author (email), January 9, 2017.

Fleming, Ian, to Mary Bond, June 20, 1961, Free Library of Philadelphia, Rare Book Department archives.
Griffies-Williams, Beryl, to Mary Bond, 1961, Free Library of Philadelphia, Rare Book Department archives.
Lack, David, to James Bond, regarding the Bond Line, which separated North and South American bird families, 1973, Free Library of Philadelphia, Rare Book Department Archives.
Lavergne, Michael, to author, December 16, 2016.
Stewart, Martha, to author (email), February 1, 2018.
Yale biology student correspondence with Jim Bond, 1965, Free Library of Philadelphia, Rare Book Department archives.

Miscellaneous

Blakely, Julia. "Bond, James Bond, the Birds, the Books, the Bonds." Unbound, the Smithsonian Libraries blog, June 12, 2016. https://blog.library.si.edu.
Briand, Frederic. "The Name Is Bond, James Bond: From Bird Scientist to Spy," *National Geographic* blog, October 12, 2012. https://blog.nationalgeographic.org.
"Ian Fleming: The Brain behind Bond." Television interview, *Explorations*, Canadian Broadcasting Corporation, broadcast August 17, 1964.
Medical certificate to Bahamas immigration officer, August 1929. Free Library of Philadelphia.

Index

Index

Page numbers in italics indicate images

A

Academy of Natural Sciences, 5, 10, 25, 33, 34, *35*, 36, 38, 39, *41*, 45, *46*, 47, *47*, 50, 51, 52, 53, *53*, 54, 55, 58, 67, 73, 82, 89, 91, 94, 96, *96*, 97, 102, 103, 108, 109, 111, 114, *115*, 135, 136, 137, 138
Acosta, Ramsey, 122, *122*
America, SS, *102*, 103, *103*, 104, *104*, 105, *105*
American Museum of Natural History, 10, 89, 90, 91, 92
Arlott, Norman, 81
Armistead, George, 107, 134
Armistead, Harry, 26, 28, 134
arsenic, 45-51, 53, 137
Ascension Island, *91*, 92, 93
Aubrecht, Gerhard, 109, 134, 137
Audubon, John James, 10, 21, 28, 31, *47*, 51, 114

B

Bahamas, 7, 34, 36, 41, 42, 43, 58, 89, 102, 139
Barbour, Thomas, 38, 135
Birds of Mt. Desert Island, Acadia National Park, Maine, 31, 62, *62*, 63, 117, 137
Birds of the West Indies, 7, 9, *9*, 10, 11, 42, 49, 52, 57-63, 65, 66, 67, 69, 70, 71, 77, 81, 84, *85*, 99, 107, 109, 120, 121, 137
Blackwell, Chris, 19, 125
Blake, Emmet Reid, 49-51, 67, 89, 98-99, *99*, 135
Blakely, Julia, 67, 139
Bond, James,
expeditions, 33-42 (*41*, *42*)
and Ian Fleming, 7-8, 13-19, *16*, 65-78
Other images, 6, *16*, 46, *53*, 60, *64*, 72, 82, *102*, 108, 144
overview, 6-11 (*11*)
youth, 20-29 (*20*, *24*, *27*, *28*)
Bond, Mary Wickham, 8, 14-18, 36, 37, 39, 41, 45, 51, 54, 58, 61, *64*, 68, 70, 71, 72, 73, 74, *74*, 75, 76, 81-84, 91, 101, 102, 107, 110, 111, 115-18, 120, 122, 124, 126, 134, 135, 138, 139
Bond's Line, 9, 10-11, 107, 108, *108*, 109, 139
Boulton, W. Rudyerd, 89, 92-93, *93*, 135
Briand, Frederic, 108, 139
Brosnan, Pierce, 11, 77, 124, *124*
Bryce, Ivar, 13, 69, 122, 135

C

Cadwalader, Charles M. B., 52, 53, 54, 55, 138
Carriker, Melbourne Armstrong, 52-55, 135, 137
Casino Royale, 7, 65, 66, *66*, 67, *67*, 83-84
Cervera, Fermín, 37-38
Chapin, James Paul, 89-93, *90*, 96, 135
Contosta, David, 21, 22, 26, 29, 52, 74, 83, 134
Cornwell, David (John Le Carré), 88
Cory, Charles B., 34, 49, 51, 57, *58*
Coues, Elliot, 48, *48*, 49, 135, 136
Crockett, Frederick, 89, 94, 97-98, 135
Cuba, 9, 11, 34, 36-39, 50, 51, 57, 101, 102, 109, 117, 119-121
Bay of Pigs, 37, 39, 101, 119, *119*
Zapata Swamp, 37-39, 119-121
curlew, Eskimo, 72-73, *72*, 138

D

Deignan, Herbert Girton, 89, 95, *95*, 96, 135

Index

De Schauensee, Rodolphe Meyer, 33, 34, 52, 53, *53*, 54, 94, 95, 98
Dolan, Brooke, II, 89, 96, *96*, 97, 102, 135
Dominican Republic, 9, 34, 36, 101, 102
Duquesne Spy Ring, 105, *105*

E

Eckelberry, Don, 58
Ekman, Erik, 36-37

F

Far Afield in the Caribbean, 36, 37, 39, 45, 74, 135
Field Ornithology, 48-49, 135
Fleming, Ian, 54, *79*, 80-85, 87, 88, 89, 101, 102, 103, 119, 121-125, 135, 136, 137, 138, 139
    and James Bond, 7-8, 13-19, (*14,16*), 65-78, 119, 121-125, 135, 136, 137, 138, 139
For Your Eyes Only, 121, *121*
Free Library of Philadelphia, 36, 82-83, 134, 138, 139
Fukuda, Hiroki, 82, 134

G

Garrido, Orlando, 59, 60, 107, 134, 136
Gill, Frank, 110, 134
Goldeneye, Jamaica, 7, 13-19, 54, 65-71, 81, 83, 121, 122, 125, 126
GoldenEye Resort, Jamaica, *12*, *122*, *123*, 124-125, 136
Goldfinger, 7, 14, 19, 71, 101
Gundlach, Juan, 107
Gwynedd Mercy University, 116, *117*, 136

H

Haiti, 9, 11, 34, 36, 37, 40, 101, 102, 103

Harrow School, 26, *26*, 27, *27*, 87, 102
Harvard, Museum of Comparative Zoology, 10
Hispaniola, 9, 11, 108
Hoatzin, 33, *35*
How 007 Got His Name, 14, 15, 54, 68, 70, 72, 73, 74, *75*, 84, 124, 135
hummingbird, 13, 67, 107, 108
    bee, 9, *9*, *50*, 119
    streamertail, 9, *10*, 121, *121*
hutia named for Bond, 10, *10*
Jamaica, 7, 9, 12-19, 34, 39, 40, *40*, 51, 65, 81, 82, 89, 102, 107, 108, 109, 111, 121-126, 135, 136

K

Kelly, William, 101
kling-klings, 13, *14*, 125
Knight, Maxwell, 87, 88, 98, 136

L

Lack, David, 9, 108, *108*, 136, 139
Levy, Catherine, 107, 134
Lewison, Grant, 109, 137

M

McCluhan, Marshall, 7
McConnell, Scott, 135
Macdonald, Helen, 87, 138
MacDonald, Rich, 63, 134
Maine, Mt. Desert Island, 24, 29, 31, 62, 63, 74, 117, 118, 137
Meinertzhagen, Richard, 49, *86*, 87, 88, 102, 135, 136

Index

N

nuthatch, Bahama, 41, 42, *43*

O

Olson, Storrs 54, 127
On Her Majesty's Ornithological Service, 73, 129-132
Owl, Snowy, *30*

P

parakeet, Carolina, *47*, 51
Parker, Matthew, 13, 65, 67, 89, 122, 136
Patrick, Ruth, 52, 111, 134
Pearson, John, 76, 101, 122, 124, 136
Peck, Robert M., 25, 82, 107, 134, 136, 137
Philadelphia, 21-29, 47, 71, 74, 94, 103, 107, 113, *113*, 114-116
Philby, Kim, 87, 102
Phillies, Philadelphia, 22-24, 138
Poole, Earl, 7, 58, 61

R

Raffaele, Herbert, 57, 59-61, 134
rail, Zapata, *39*
Rice, Nate, 51, 53, 109, 134
Ripley, S. Dillon, II, 26, 89, 94, *94*, 95, 98, 102, 108, 135, 136

S

Simon, Taryn, 57, 61-62
sparrow, Zapata, 38, 120-121
St. Paul's School, 26, 94, 96, 102, 136
Stone, Witmer, 34, 36, 52, 91, 136, 138

T

Thomson, Keith, 109, 111, 134
To James Bond with Love, 41, 45, 51, 68, 73, 74, 76, 83, 91, 101, 102, 103, 110, 124, 126, 135
tody, Cuban, 57, 58, 61, *120*, 121
Trinity College, 27, 28, 102, 116
Turvey, Samuel, 10-11, 134
Tyson, Carroll Sargent, Jr., *21*, 22, *29*, 29-31, 62-63, 117, 118, 134
Tyson, Charles, Jr., 30-31, 134

U

unicorn bird, *53*, 54

W

wideawakes, *91*, 92
Willow Brook, 25, 26, 115-116, *117*, 137
Wren, Zapata, 38, 120-121

Y

Yale University, Peabody Museum of Natural History, 10, 89, 95, 109
You Only Live Twice, 14, 17, 19, *80*, 81-84, 85, 136

Z

Zapata Swamp, 37, 38, *38*, 39, 120-121